Engineering Materials

This series provides topical information on innovative, structural and functional materials and composites with applications in optical, electrical, mechanical, civil, aeronautical, medical, bio- and nano-engineering. The individual volumes are complete, comprehensive monographs covering the structure, properties, manufacturing process and applications of these materials. This multidisciplinary series is devoted to professionals, students and all those interested in the latest developments in the Materials Science field, that look for a carefully selected collection of high quality review articles on their respective field of expertise.

Indexed at Compendex (2021)

Xinying Lu

Passivation and Corrosion of Black Rebar with Mill Scale

Springer

Xinying Lu
School of Civil Engineering
Tsinghua University
Beijing, China

ISSN 1612-1317 ISSN 1868-1212 (electronic)
Engineering Materials
ISBN 978-981-19-8101-2 ISBN 978-981-19-8102-9 (eBook)
https://doi.org/10.1007/978-981-19-8102-9

This Springer imprint is published by the registered company Springer Nature Singapore Pte Ltd.
The registered company address is: 152 Beach Road, #21-01/04 Gateway East, Singapore 189721, Singapore

Doing research is somewhat like the
"blind men and an elephant".
Life journey is the same...

Preface

Corrosion of black rebar is an old topic. There are so many publications available, and the causes either by concrete carbonation or chloride attack are well known. It seems nothing new or worthy to talk.

After so many rebar corrosion cases encountered in the past decades, I got realized that there might be some big differences in corrosion behavior between the black rebar with mill scale in field and the polished one in lab. So, we began to work on it in the late 1990s.

After the millennium, Prof. Shiyuan Huang from Tongji University encouraged me to write an elementary book on rebar corrosion for civil engineers based on our observations to reduce the problems in practice, and gave me Dr. Tuutti's dissertation as encouragement.

Unfortunately, the expected book never comes out since we haven't got the final answer even in the last 20 years.

In order not to regret more, I have prepared this monograph to provide some of our preliminary observations for engineers' reference even though it is far from perfect. The readers are assumed to have some basic knowledge about the metallic corrosion already. It includes nine chapters:

Chapter 1: Passivation of iron. Mainly introduces the discovery of passivation and the passive films of pure iron under different conditions.

Chapter 2: Determination methods of passivation. Potentiodynamic polarization and surface analysis methods are briefly introduced.

Chapter 3: Passivation of rebars without mill scale. Mainly introduces the passive film of rebar without mill scale under strong alkaline conditions, which is the foundation of the existing principles.

Chapter 4: Mill scale of hot-rolled rebars. Mainly introduces the formation, composition, and structure of the mill scale as well as its semiconductive characteristics.

Chapter 5: Passivation of hot-rolled rebars. The main reaction mechanisms and passivation behavior of rebar with mill scale are introduced.

Chapter 6: Redox reactions of hot-rolled rebars. Mainly introduces the redox phenomenon and the mechanisms of the mill scale.

Chapter 7: Corrosion of hot-rolled rebars. The differences in corrosion between hot-rolled rebars and the descaled ones are briefly introduced.

Chapter 8: Protection of hot-rolled rebars. Briefly introduces some measures that can be used for the corrosion prevention of hot-rolled rebars.

Chapter 9: Further work needs to do. Some considerations about the future work.

Here, I would like to say thanks to my students, who once worked on this subject or some work related, Wei Lin, Yuanjin Li, Xin Wang, Guiyang Zhang, Kun Jiang, Xinyun Gu, Xiaojia Liu, Hao Li, Rongpeng Li, Yang Li, Chao Lin, Fei Lin, Xiaoshuang Wang, etc. This monograph mainly quotes the key results of the first three students. Especially thank Yuanjin Li and Xin Wang for their primary contributions to this monograph.

Thanks also to my teachers, friends, and relatives who have always encouraged, supported, and cared about me!

I'd like to say thanks to those friends who have given the criticism and negative opinions. Without your help, we would take much more detours. Thank you very much!

Limited to our knowledge, mistakes in the monograph are inevitable. Any comments or corrections are all welcome.

Beijing, China
February 2022

Xinying Lu

About This Book

The passivation and corrosion of any metal or alloy are significantly affected by its surface state and chemical characteristics. Therefore, the passivating and corroding behaviors of the black rebar with mill scale or rust stains are quite different from that of the descaled one. The pseudo-passivation and the corrosion mechanisms as well as the corresponding protection measures of the hot-rolled black rebars are briefly introduced in this monograph. The early pre-rusted rebars are ready to corrode all the time and they need more attention in practice. This monograph can be a reference for corrosion or civil engineers.

Contents

Chapter 1
Passivation of Iron

Abstract The passivation phenomenon and its explanation, the passive films and their structures of pure iron are briefly reviewed. The nano-thick passive film of Fe is usually composed of an inner layer of Fe^{2+} compound and an outer layer of Fe^{3+} compound, such as a $Fe_3O_4/\gamma\text{-}Fe_2O_3$ or $\alpha\text{-}Fe_2O_3$ bilayer.

1.1 Discovery and Explanation of Passivation

As shown in Fig. 1.1, if a piece of pure Fe is put into the dilute HNO_3, it will dissolve and H_2 bubbles release from the solution. However, if it is put into the concentrated HNO_3, there is no reaction or after a very slight reaction with a minor amount of gas released, then it stops immediately.

If dip the Fe in the concentrated HNO_3 first, and put it back into the dilute HNO_3 again, then the dissolution of Fe will not happen immediately or not at all.

The dissolution of Fe in dilute HNO_3 is called corrosion or activation, while the insoluble phenomenon in concentrated HNO_3 is called passivation.

Metallic corrosion is the phenomenon that a metal interacts with the environment, loses electrons, and is oxidized (see Eq. 1.1*). The metal may lose weight or its properties get degraded.*[1]

$$M - ne \rightarrow M^{n+} \tag{1.1}$$

The history of research on iron passivation is almost near 300 years, and it can be roughly divided into three stages:

(1) eighteenth century, the passivation phenomenon was discovered;
(2) nineteenth century, the passivation mechanisms were explained and widely debated;
(3) since the twentieth century up to now, the passivation mechanisms have been verified and the passive films structures have been deeply analyzed, see Fig. 1.2.

[1] The content in italics is the related or extended knowledge.

X. Lu, *Passivation and Corrosion of Black Rebar with Mill Scale*, Engineering Materials,
https://doi.org/10.1007/978-981-19-8102-9_1

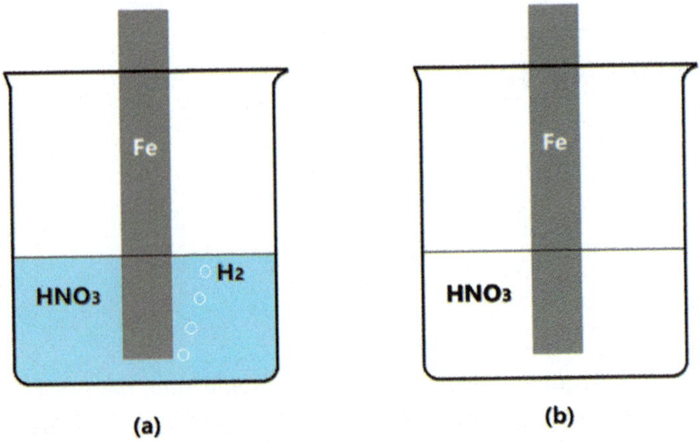

Fig. 1.1 Reaction of Fe in HNO_3, (a) corrosion in dilute HNO_3 and (b) passivation in concentrated HNO_3

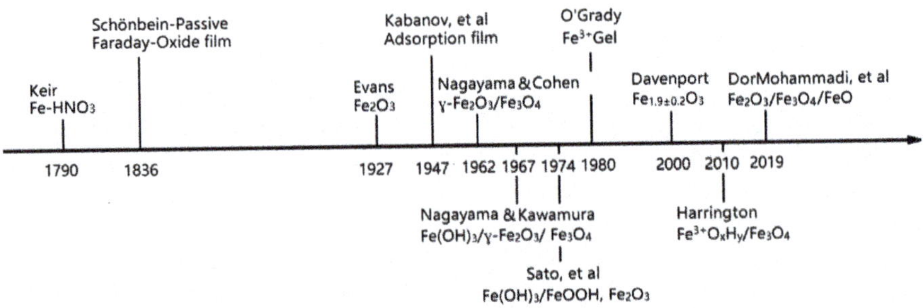

Fig. 1.2 Schematic diagram of iron passivation research history

It is said that as early as 1738, Russian chemist Lomonosov discovered the passivation of Fe in concentrated nitric acid [1]. In 1790, Keir of Scotland found that Fe did not react with the concentrated nitric acid nor the concentrated sulfuric acid at room temperature, and the pre-immersed Fe in concentrated HNO_3 could not convert Ag from the $AgNO_3$ solution anymore [2].

During 1830s, British scientist Faraday and German-Swiss chemist Schönbein studied the reactions of Fe in different concentrations of HNO_3 [3, 4]. And in 1836, Schönbein first coined the behavior of Fe in concentrated HNO_3 as "passive". In the same year, Faraday said in his reply to Schönbein [3], the reason why Fe can passivate in concentrated HNO_3 is mainly due to the oxidation of Fe and the formation of an oxide film on its surface, which is the famous theory of "oxide film formation" or "oxide film".

However, Schönbein did not agree with this statement [4]. He believed that the passive Fe was caused by the negative catalysis of concentrated HNO_3.

In the following hundred years, many people studied the passivation of iron and put forward a variety of explanations. It was not until 1927 that Evans successfully

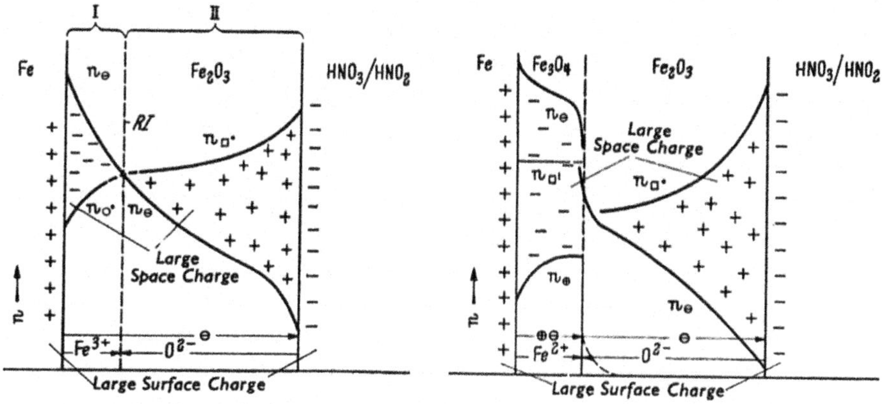

Fig. 1.3 Passive film formation under electric field or potential gradient, reprinted from [7] with permission of Springer Nature

peeled off the passive films on the Fe surface by electrolysis or KI solution dissolution [5] that Faraday's statement was verified.

Influenced by Langmuir's adsorption theory, in 1919, Tammann proposed that the passive film of Fe could be a layer of oxygen atom adsorption film. In 1958, Uhlig proposed that the passive film of Fe could be a layer of oxygen atom or oxygen molecule adsorption film, the oxygen atom could be provided by OH or H_2O in the solution [6], and he believed that Faraday's statement had included this viewpoint, which is the famous theory of "adsorption film".

In twentieth century, with the fast developments of those advanced electrochemical and surface analysis instruments, people have widely studied the passivation behavior of Fe in different solutions under different conditions, and put forward a variety of reaction mechanisms, such as the high electric field or its modified models [7, 8] (see Fig. 1.3 for example), the deposit film model [9, 10] (see Fig. 1.4), and the Point Defect Model (PDM) [11] (see Fig. 1.5), etc.

Unfortunately, although so much work has been done, so far, there is no unified conclusion has been reached on the passivation mechanisms and the passive film compositions of pure iron yet.

1.2 E-pH

Obviously, the active and passive processes of Fe in different pH aqueous solutions are different.

Figure 1.6 shows the E-pH diagram when considering iron oxides formed in aqueous solution [12]. In different potential and pH zones, Fe can demonstrate three

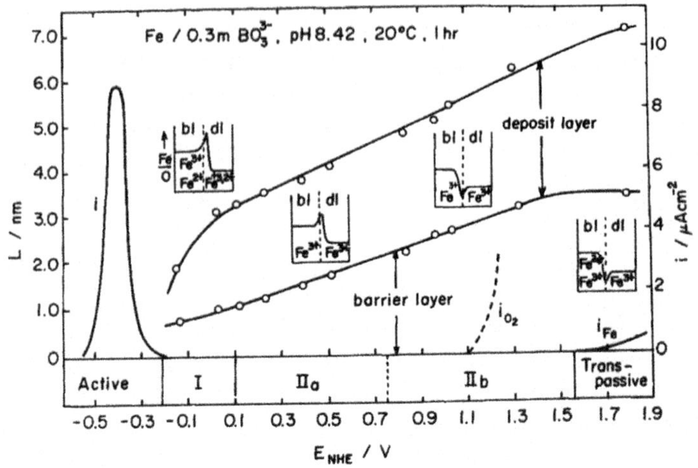

Fig. 1.4 The barrier-deposit bilayer structure of passive film, reprinted from [9] with permission of Springer Nature

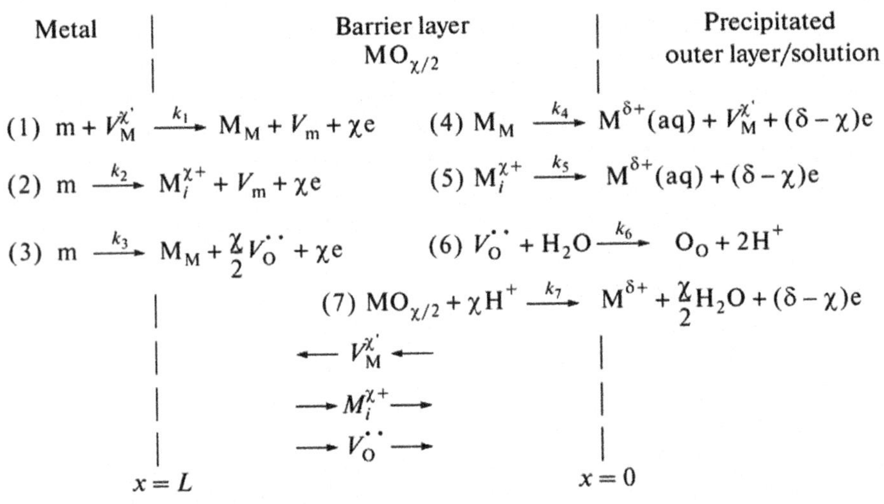

Fig. 1.5 The PDM model, reprinted from [11] with permission of Springer Nature. $V_M{}^{\chi'}$ cation vacancy, $M_i{}^{\chi+}$ cation interstitial, $V_O^{\cdot\cdot}$ oxygen vacancy, $M^{\delta+}$ (aq) cation in outer layer/solution interface, M_M cation in cation site on the cation sublattice, O_O oxide ion in anion site on the anion sublattice, $MO_{\chi/2}$ stoichiometric barrier layer oxide. The reactions (1), (2), (4), (5), and (6) are lattice conservative processes (they do not result in the movement of the interface), whereas reactions (3) and (7) are non-conservative

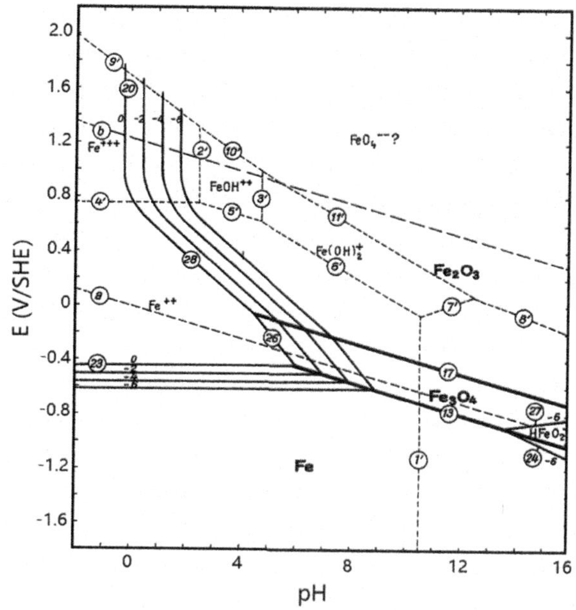

Fig. 1.6 E-pH diagram of Fe-H_2O at 25 °C, 1 atm, considering Fe_xO_y generated, reprinted from [12] with permission of Elsevier

different states: immune (Fe), corrosive (Fe^{2+}, Fe^{3+}, $HFeO_2^-$), and passive (Fe_3O_4, Fe_2O_3).

According to Fig. 1.6, in the weak acidic and neutral solutions, in order to passivate Fe, its potential must be increased. With the increase of pH, the required potential to passivate is decreased.

In a strong alkaline solution, such as pH=12, Fe can generate Fe_3O_4 or Fe_2O_3 at potentials greater than −0.5 or 0 V/SHE,[2] respectively. If the oxides cover the metal surface continuously and densely, they may inhibit the substrate further corrosion (or reduce the corrosion rate to a very low level), i.e., the metal gets passivated.

When pH > 13, however, Fe can generate $HFeO_2^-$ and begin to corrode, that is, Fe cannot maintain chemical stability in very strong alkaline conditions.

If iron is oxidized to hydroxides, then the E-pH diagram of Fe-H_2O system will change to Fig. 1.7. When the hydroxides cannot deposit on the Fe surface and the dissolution process cannot be inhibited, then iron is not able to be passivated.

1.3 Passive Films of Iron

In 1959, Cohen [13] analyzed the passivation of Fe under different conditions (see Table 1.1), trying to give a unified explanation.

---

[2] SHE: Standard Hydrogen Electrode.

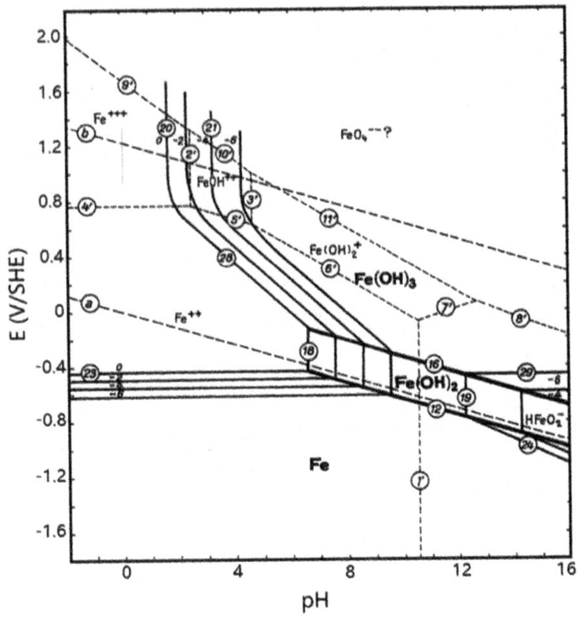

Fig. 1.7 E-pH diagram of $Fe-H_2O$ at 25 °C, 1 atm, considering $Fe(OH)_x$ generated, reprinted from [12] with permission of Elsevier

Cohen [13] believed that the passive film of Fe is mainly composed of $\gamma-Fe_2O_3$ and considered $\gamma-Fe_2O_3$ is chemically inert and insoluble in acid, but it can be dissolved by cathodic reduction to Fe^{2+} in neutral or acidic solution.

Although the passive film of Fe is a good electronic conductor, it is not an ionic conductor. Therefore, the growth of the passive film is mainly controlled by the ion transfer rate.

There are usually pores or local discontinuities in the passive film, and some iron hydroxides or other insoluble salts are embedded. When water or aggressive ions accumulate at these defects, it will lead to local acidification or formation of soluble complex salts, which can cause local damage of the passive film.

Since the X-ray diffraction patterns of $\gamma-Fe_2O_3$ and Fe_3O_4 are very similar, it can be considered that $\gamma-Fe_2O_3$ is somewhat the Fe_3O_4 with cation defects. However, it can be inferred from the different thicknesses of the passive films that there are obvious differences in the diffusion characteristics of cations between them. In addition, their chemical properties are also different: $\gamma-Fe_2O_3$ is insoluble in acid, while Fe_3O_4 is soluble; $\gamma-Fe_2O_3$ can be completely reduced to Fe^{2+} ions by cathodic polarization, while the reduction of Fe_3O_4 is relatively difficult. From the increase of lattice constant and the difference in X-ray diffraction patterns, it can be deduced that Fe_3O_4 can transform into $\gamma-Fe_2O_3$ and partial recrystallization may occur in such a process.

The different growth rate and final film thickness of the two oxides may be due to the different migration rates of Fe^{2+} at the metal/oxide interface. To some extent, it is also related to their different crystal arrangements on the metal surface. Fe_3O_4 films usually show preferred orientation on Fe surface, while $\gamma-Fe_2O_3$ film does not.

Table 1.1 Passivation and reaction mechanisms of Fe under different conditions, reprinted from [13] with permission of Canadian Science Publishing

Passivating conditions		Reaction mechanisms
1. Reaction of Fe with H_2O		
(1)	Absence of oxygen	Generates $Fe(OH)_2$ or Fe_3O_4
		1) $Fe \rightarrow Fe^{2+} + 2e \Rightarrow 2H_2O + 2e \rightarrow 2OH^- + H_2$ 2) $Fe + 2H_2O \rightarrow Fe(OH)_2 + H_2$ 3) $3Fe + 4H_2O \rightarrow Fe_3O_4 + 4H_2$ (When temperature rises, or other chemicals exist)
		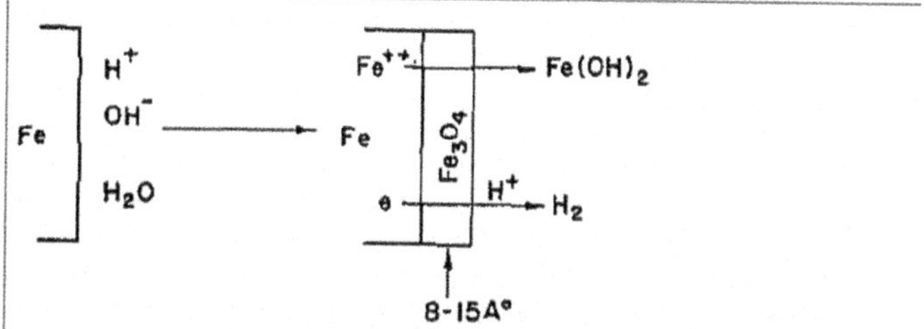
(2)	Absence of oxygen but presence of oxidizing ions	1) Low concentration nitrite: generates Fe_3O_4; High concentration nitrite: generates γ-Fe_2O_3 or γ-FeOOH 2) Chromate: generates γ-$Fe_2O_3 + Cr_2O_3$, 50–75 Å 3) Molybdate or tungstate: generates Fe_3O_4; Anodic oxidation forms γ-Fe_2O_3
		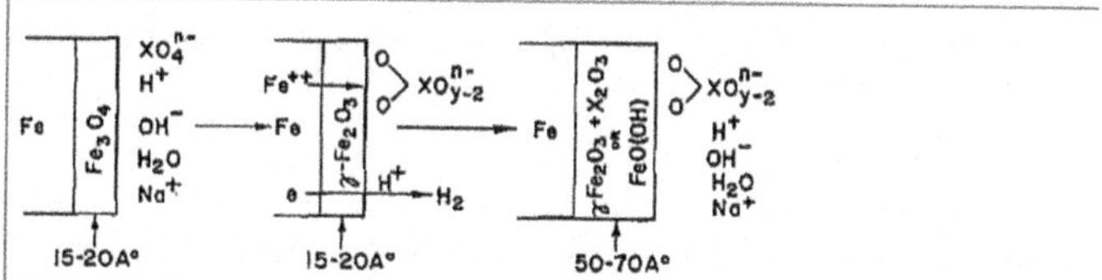
(3)	Presence of oxygen	1) Dry air: generates γ-Fe_2O_3, 15–20 Å 2) Normal temperature, oxygen-enriched and humid conditions: forms $Fe(OH)_3$ 3) If the metal surface is covered with iron hydroxide, Fe_3O_4 and iron hydroxide deposition layers will be formed 4) Fe_3O_4 can be oxidized to γ-Fe_2O_3; Fe^{2+} diffused to the solution interface can be oxidized to γ-FeOOH 5) The deposition layer will promote the formation of Fe_3O_4 at low oxygen concentrations and γ-Fe_2O_3 formation at high oxygen concentrations
(4)	Presence of oxygen and oxidizing ions	1) When there is oxygen, it will reduce the concentration of oxidizing anions required for passivation 2) With oxidizing anions, the thickness of passive film can reach 50-75 Å 3) In nitrite, molybdate, or tungstate solution, oxygen can promote the conversion of Fe_3O_4 into γ-Fe_2O_3

(continued)

Table 1.1 (continued)

Passivating conditions		Reaction mechanisms
(5)	Presence of oxygen and non-oxidizing ions	Na_2CO_3, Na_3PO_4, or C_6H_5COONa, forming γ-Fe_2O_3, iron hydroxide may embed in oxide films
2. Reactions in presence of air-formed film		
(1)	With passivators	Air generated γ-Fe_2O_3 layer will be further thickened and iron hydroxide may be formed
(2)	At critical passivating conditions	Partial γ-Fe_2O_3 may be reduced, new oxide film embedded
3. Effect of applied current		
(1)	Cathodic current	At low current density, γ-Fe_2O_3 will be reduced to Fe_3O_4 At high current density, it is possible to be reduced to Fe
(2)	Anodic current	Corrosion may occur at low current density and passivation may occur at high current density

Cohen [14] summarized six compositions and structural diagrams of Fe passive films, as shown in Fig. 1.8. Among them, A is the most common "bilayer" oxide structure, but there is not a clear interface necessarily between the two layers, and the concentration of Fe decreases gradually from inside to outside (or oxygen concentration from outside to inside). B is the case where the outer layer of the passive film contains hydrogen protons or water molecules. C is the monolayer of γ-Fe_2O_3 film presented in the acidic solution. D is the formation of an outer layer of γ-FeOOH deposition film while Fe^{2+} is dissolved in the neutral solution. E indicates that the passive film always contains hydrogen or water, but the oxide film is usually seen as dehydrated. F is the typical passive film structure of Fe in neutral borate solution. The outer layer is metal deficient iron oxide, which does not exclude hydrogen protons. The inner layer is Fe_2O_3 and Fe_3O_4.

Nishimura and Sato [10] studied the passive behavior of Fe in borate and phosphate solutions with different pH values. They found that:

(1) When $pH < 2$, the structure near the Fe surface is Fe|Fe_3O_4 inner film|γ-Fe_2O_3 outer film|solution;
(2) When $2 < pH < 5.5$, the structure near the Fe surface is Fe|Fe_3O_4 inner film|γ-Fe_2O_3 outer film| $Fe(OH)_3$ deposition film|solution;
(3) When $pH > 5.5$, the structure near the Fe surface is Fe|Fe_3O_4 inner film |$Fe(OH)_3$ deposition film|solution.

McCafferty [15] classified the composition and structure of passive films on pure Fe and Fe-based alloys into four types: bilayer oxide film, hydroxide film, spinel-like film, and bipolar ion-selective film. The first three have been mentioned above. The fourth is composed of an outer positive ion-selective film and an inner negative ion-selective film. The outer layer only allows cations to pass through and the inner

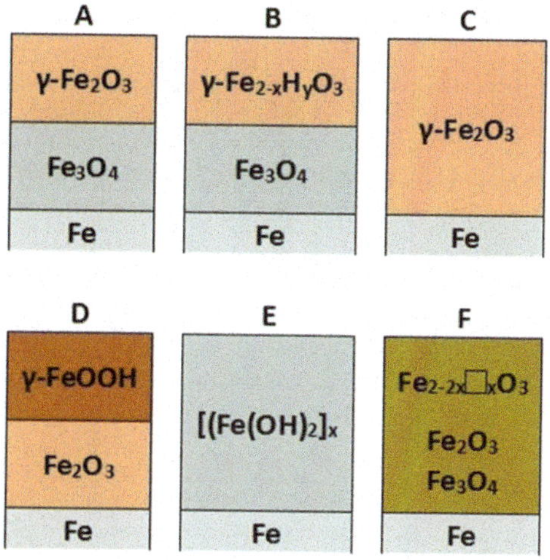

Fig. 1.8 Composition and structure of Fe passive films, reprinted from [14] with permission of IOP Publishing

layer only allows anions to pass through, which can inhibit the oxidation of the Fe matrix.

Cahan and Chen [16] believed that the passive film of Fe is a kind of "chemical conductor film", which is a wide range, non-stoichiometric oxide or hydroxide film composed of low valent iron ions rich near the Fe interface and high valent iron ions rich near the solution interface.

Davenport et al. [17] considered that the passive film of Fe formed in weak alkaline borate solution is neither Fe_3O_4 crystal phase nor γ-Fe_2O_3 crystal phase but a new phase, which is named LAMM. This phase has a spinel structure. The oxygen ion crystal position is completely occupied by oxygen, while the octahedral occupation of iron is $80 \pm 10\%$, the tetrahedral occupation is $66 \pm 10\%$, and the octahedral interstitial occupation is $12 \pm 4\%$. That is, the composition of the passive film is similar to Fe_3O_4 or γ-Fe_2O_3, but the occupancy of Fe in octahedron, tetrahedron, and interstitial sites is obviously different, which is a nanocrystalline microstructure with a large number of extended defects (inverse boundary and stacking fault).

So far, there are many different points on the composition, structure, thickness, compactness, and conductivity of Fe passive films, see Table 1.2.[3] Generally, the passive film of Fe is about several angstroms to tens of nanometers thick, which is usually divided into two layers, the inner is mainly Fe^{2+} compound and the outer is mainly Fe^{3+} compound (which can be oxide, hydroxide, or hydroxyl oxide). There are local defects or discontinuities in the film, and the film is a good conductor of electrons.

[3] The passive film components in the table are arranged from the outside to the inside. Chem-Chemical, E-Chem-Electrochemical, Opt-Optical, TEM-Transmission Electron Microscope, AES-Auger Electron Spectroscopy, XPS-X-ray Photoelectron Spectroscopy, SIMS-Secondary Ion Mass Spectroscopy, Raman-Raman Spectra, STM-Scanning Tunneling Microscope, XANES-X-ray Absorption Near Edge Structure, SER-Surface Enhanced Raman Scattering, XRD-X-ray diffraction, ReaxFF-Molecular Dynamics Simulation, for abbreviations.

Table 1.2 Preparation method, composition, structure, and passivation model of passive films of iron

Time	Researcher	Passivating media	pH	Preparation	Analysis method	Comp. & Struc.	Thick-ness (nm)	Model	Ref.
1790	Keir	$H_2SO_4/HNO_2/KNO_3$	/	Chem	Visual	/	/	Oxide film	[2]
1836	Schönbein and Faraday	HNO_3	/	Chem	Visual	Oxide	/	Oxide film Oxygen adsorption film	[4, 18]
1911	Dunstan and Hill	$K_2Cr_2O_7/H_2CrO_4/HNO_3$; O_2	/	Chem	Visual	Fe_3O_4	/	Oxide film	[19]
1927	Evans	$K_2Cr_2O_7/KNO_3/NaOH/HNO_3$; O_2	/	Chem	KI dissolution or electrolysis	Fe_2O_3	/	Oxide film	[5]
1931	Müller	H_2SO_4/Na_2SO_4	/	E-Chem	E-Chem /Opt	Oxide/deposition film	/	Oxide film	[20]
1933	Tronstad	$NaOH/Na_2SO_4/H_2SO_4$	/	E-Chem	Ellipsometry	Oxide	2.0–4.0	Oxide film	[21]
1947	Kabanov et al.	$KOH/NaOH/Na_2SO_4/NaCl$	/	E-Chem	E-Chem	Adsorption film	/	OH^- adsorption film	[22]
1962	Nagayama and Cohen	$Na_2B_4O_7$-H_3BO_3	8.41	E-Chem	E-Chem	γ-Fe_2O_3/Fe_3O_4	1.5–3.0	Bi-oxide films	[23]
1967	Nagayama and Kawamura	$Na_2B_4O_7$-H_3BO_3	8.43 6.50	E-Chem	E-Chem, Radiochemistry	$Fe(OH)_3$/ γ-Fe_2O_3/ Fe_3O_4	2.0–8.0	Oxide film	[24]
1967	Foley et al.	$NaOH/Na_2B_4O_7$-H_3BO_3/H_2SO_4	11.4 8.5 1.1	E-Chem	TEM	γ-Fe_2O_3 Fe_3O_4 γ-$FeOOH$	~3.0	Oxide film	[25]
1970	Sato et al.	$Na_2B_4O_7$-H_3BO_3	8.4	E-Chem	E-Chem/ Opt	$Fe_2O_3{\cdot}0.39H_2O$	1.3–1.5	Oxide film	[26]

(continued)

Table 1.2 (continued)

Time	Researcher	Passivating media	pH	Preparation	Analysis method	Comp. & Struc.	Thick-ness (nm)	Model	Ref.
1974	Sato et al.	$Na_2B_4O_7$-H_3BO_3/NaH_2PO_4-Na_2HPO_4/NaOH/H_2SO_4	11.50 ~1.45	E-Chem	E-Chem/Opt	$Fe(OH)_3$ FeOOH Fe_2O_3	1.5–4.5	Multi-oxide films	[27]
1975	Revie et al.	KOH-H_3BO_3	8.1	E-Chem	AES	$Fe_2O_3 \cdot H_2O$	/	Oxide film	[28]
1976	Ord and Smet	$Na_2B_4O_7$-H_3BO_3	8.4	E-Chem	E-Chem/Opt	Fe_2O_3/Fe_3O_4	/	Bi-oxide films	[29]
1977	Seo et al.	KOH-H_3BO_3	8.4	E-Chem	AES	Fe_2O_3/Fe_3O_4	~3.0	Bi-oxide films	[30]
1980	O'Grady	$Na_2B_4O_7$-H_3BO_3	8.4	E-Chem	Mössbauer	Fe^{3+}Gel	/	Amorphous	[31]
1982	Kuroda et al.	$Na_2B_4O_7$-H_3BO_3	8.4	E-Chem	TEM	Fe_2O_3/ Fe_3O_4	/	Spinel	[32]
1982	Eldridge et al.	$Na_2B_4O_7$-H_3BO_3	8.4	E-Chem	Mössbauer	Fe^{3+}Gel	/	Amorphous	[33]
1982	Chen and Cahan	$Na_2B_4O_7$-H_3BO_3	8.4	E-Chem	E-Chem/Opt	$Fe^{3+}O_xH_y$	/	Oxide film	[34, 34]
1985	Zakroczymski et al.	NaOH	12	E-Chem	E-Chem/Opt	FeOOH	~5	Oxide film	[36, 37]
1986	Brett et al.	$Na_2B_4O_7$-H_3BO_3	8.4	E-Chem	Mössbauer XPS	γ-Fe_2O_3 γ-FeOOH	/	Oxide film	[38]
1987	Haupt and Strehblow	NaOH	/	E-Chem	XPS	$Fe^{3+}O_x/Fe^{2+}O_y$	1–3	Bi-oxide films	[39]
1988	Bardwell et al.	$Na_2B_4O_7$-H_3BO_3/Na_2SO_4/H_2SO_4	8.4 3–5 0.27	E-Chem	SIMS	γ-Fe_2O_3 γ-Fe_2O_3/Fe_3O_4	1–3	Bi-oxide films	[40]
1989	Boucherit et al.	KOH/H_2SO_4	/	E-Chem	Raman	α-FeOOH	~2.8	Oxide film	[41]

(continued)

Table 1.2 (continued)

Time	Researcher	Passivating media	pH	Preparation	Analysis method	Comp. & Struc.	Thick-ness (nm)	Model	Ref.
1989	Larramona and Gutiérrez	$NaOH/Na_2B_4O_7$-H_3BO_3/H_2SO_4	13/14 9.2 1.0	E-Chem	E-Chem/ Opt	FeOOH Fe^{2+}	/	Single or bi-oxide film	[42]
1991	Gui and Devine	$Na_2B_4O_7$-H_3BO_3	8.4	E-Chem	Raman	$Fe^{2+}Fe^{3+}(OH)_x$	/	Single or bi-oxide film	[43, 44]
1995	Ryan et al.	$Na_2B_4O_7$-H_3BO_3	8.4	E-Chem	STM	γ-Fe_2O_3/Fe_3O_4	/	Bi-oxide films	[45]
1995	Davenport and Sansone	$Na_2B_4O_7$-H_3BO_3	8.4	E-Chem	XANES	γ-Fe_2O_3-Fe_3O_4	/	Compound oxide films	[46]
1999	Schroeder and Devine	$Na_2B_4O_7$-H_3BO_3	8.4	E-Chem	SER	γ-Fe_2O_3/Fe_3O_4	/	Bi-oxide films	[47]
2000	Davenport et al.	$Na_2B_4O_7$-H_3BO_3	8.4	E-Chem	In situ surface XRD	$Fe_{1.9\pm0.2}O_3$ (LAMM)	/	Spinel like Oxide film	[17]
2001	Joiret et al.	NaOH	14	E-Chem	Raman	FeOOH Fe_3O_4	/	Bi-oxide films	[48]
2005	Sato et al.	$Na_2B_4O_7$-H_3BO_3	8.4	E-Chem	XRD	Fe_3O_4-like	3	Spinel like Oxide film	[49]
2007	Deng et al.	$Na_2B_4O_7$-H_3BO_3	8.4	E-Chem	AES, STM	γ-Fe_2O_3-like	/	Spinel like Oxide film	[50]
2010	Harrington et al.	$Na_2B_4O_7$-H_3BO_3	8.4	E-Chem	SER	$Fe^{3+}O_xH_y/Fe_3O_4$	/	Bi-oxide films	[51]
2019	DorMohammadi et al.	NaOH	13.5	/	ReaxFF-MD, XPS	$Fe_2O_3/Fe_3O_4/FeO$	/	Tri-oxide films	[52]

1.4 Summary

A brief history about the research of the pure iron passivation is presented. Although there is no unified statement on the composition and structure of the passive film of pure iron yet, most people are willing to admit that the passive film of pure iron is a nano-oxide film with a bilayer structure. The inner layer is mainly composed of Fe^{2+}oxide, and the outer is mainly of Fe^{3+} oxide. Between the two layers, there may be not a clear interface existed.

References

1. Tomashov, N.D., Chernova, G.P.: Passivity and Protection of Metals Against Corrosion. Plenum Press, New York (1967)
2. Keir, J.: Experiments and observations on the dissolutions of metals in acids, and their precipitations; with an account of a new acid compound menstruum, useful in some technical operations of parting metals. Phil. Trans. Roy. Soc. London. **80**, 359–384 (1790)
3. Meldola, R.: The letters of faraday, and schönbein, 1836–1862; with notes, comments and references to contemporary letters. Nature **61**, 337–340 (1900)
4. Schönbein, C.F.: XXXVI. Remarks on faraday's hypothesis with regard to the causes of the neutrality of iron in nitric acid. Philos. Mag. Ser. 3 **10**(60), 172–174 (1837)
5. Evans, U.R.: CXL.-The passivity of metals. Part I. The isolation of the protective film. J. Chem. Soc. (London) **127**, 1020–1040 (1927)
6. Uhlig, H.H.: The adsorption theory of passivity and the flade potential. Zeitschrift für Elektrochemie. **62**(6–7), 626–632 (1958)
7. Hauffe, K.: Scaling processes in metals and alloys with formation of thick protective layers. In: Oxidation of Metals, pp. 417–422. Springer, Boston, MA (1965)
8. Vetter, K.J.: General kinetics of passive layers on metals. Electrochim. Acta **16**(11), 1923–1937 (1971)
9. Sato, N., Okamoto, G.: Electrochemical passivation of metals. In: Bockris, J.O'M., Conway B.E., Yeager, E., et al. (eds.)Electrochemical Materials Science. Comprehensive Treatise of Electrochemistry, Vol. 4, pp. 193–245. Boston, Springer (1981)
10. Nishimura, R., Sato, N.: Potential-pH diagram of composition/structure of passive films on iron. J. Jpn. Inst. Metals **47**(12), 1086–1093 (1983)
11. Macdonald, D.D.: Some personal adventures in passivity-A review of the point defect model for film growth. Russ. J. Electrochem. **48**, 235–258 (2012)
12. Pourbaix, M.: Atlas of electrochemical equilibria in aqueous solutions. In: Franklin, J.A. (Trans.), pp. 312–313. Pergamon, Oxford (1966)
13. Cohen, M.: The formation and properties of passive films on iron. Can. J. Chem. **37**(1), 286–291 (1959)
14. Cohen, M.: The passivity and breakdown of passivity on iron. In: Frankenthal, R.P., Kruger, D.J. (eds.) Passivity of Metals, pp. 521–545. The Electrochemical Society, Princeton (1978)
15. McCafferty, E.: Introduction to Corrosion Science, pp. 209–262. Springer, New York (2010)
16. Cahan, B.D., Chen, C.T.: The nature of the passive film on iron: III. The chemi-conductor model and further supporting evidence. J. Electrochem. Soc. **129**(5), 921–925 (1982)
17. Davenport, A.J., Oblonsky, L.J., Ryan, M.P., et al.: The structure of the passive film that forms on iron in aqueous environments. J. Electrochem. Soc. **147**(6), 2162–2173 (2000)
18. Faraday, M.: Experimental Researches in Electricity. J. M. Dent & Sons Ltd, New York (1922)
19. Dunstan, W.R., Hill, J.R.: CCIX—The passivity of iron and certain other metals. J. Chem. Soc. Trans. **99**, 1853–1866 (1911)

20. Müller, W.J.: On the passivity of metals. Trans. Faraday Soc. **27**, 737–751 (1931)
21. Tronstad, L.: The investigation of thin surface films on metals by means of reflected polarized light. Trans. Faraday Soc. **29**, 502–514 (1933)
22. Kabanov, B., Burstein, R., Frumkin, A.: Kinetics of electrode processes on the iron electrode. Discuss. Faraday Soc. **1**, 259–269 (1947)
23. Nagayama, M., Cohen, M.: The anodic oxidation of iron in a neutral solution: I. The nature and composition of the passive film. J. Electrochem. Soc. **109**, 781–790 (1962)
24. Nagayama, M., Kawamura, S.: Anodic oxidation of ferrous ion on passive iron. Electrochim. Acta **12**(8), 1109–1119 (1967)
25. Foley, C.L., Kruger, J., Bechtoldt, C.J.: Electron diffraction studies of active, passive, and transpassive oxide films formed on iron. J. Electrochem. Soc. **114**(10), 994–1001 (1967)
26. Sato, N., Kudo, K., Noda, T.: Single layer of the passive film on Fe. Corros. Sci. **10**, 785–794 (1970)
27. Sato, N., Noda, T., Kudo, K.: Thickness and structure of passive films on iron in acidic and basic solution. Electrochim. Acta **19**(8), 471–475 (1974)
28. Revie, R.W., Baker, B.G., Bockris, J.O.M.: The passive film on iron: an application of auger electron spectroscopy. J. Electrochem. Soc. **122**(11), 1460–1466 (1975)
29. Ord, J.L., De Smet, D.J.: The anodic oxidation of iron: overpotential analysis for a two-phase film. J. Electrochem. Soc. **123**(12), 1876–1882 (1976)
30. Seo, M., Sato, M., Lumsden, J.B., et al.: Auger analysis of the anodic oxide film on iron in neutral solution. Corros. Sci. **17**(3), 209–217 (1977)
31. O'Grady, W.E.: Mössbauer study of the passive oxide film on iron. J. Electrochem. Soc. **127**(3), 555–563 (1980)
32. Kuroda, K., Cahan, B.D., Nazri, G.H., et al.: Electron diffraction study of the passive film on iron. J. Electrochem. Soc. **129**(10), 2163–2169 (1982)
33. Eldridge, J., Kordesch, M.E., Hoffman, R.W.: In situ mössbauer studies of passive films on iron. J. Vac. Sci. Technol. **20**(4), 934–938 (1982)
34. Chen, C.T., Cahan, B.D.: The nature of the passive film on iron: I. Automatic ellipsometric spectroscopy studies. J. Electrochem. Soc. **129**(1), 17–26 (1982)
35. Cahan, B.D., Chen, C.T.: The nature of the passive film on iron: II. A-C Impedance Studies. J. Electrochem. Soc. **129**(3), 474–480 (1982)
36. Zakroczymski, T., Fan C.J., Szklarska-Smialowska, Z.: Kinetics and mechanism of passive film formation on iron in 0.05M NaOH. J. Electrochem. Soc. **132**(12), 2862–2867 (1985)
37. Zakroczymski, T., Fan, C.J., Szklarska-Smialowska, Z.: Passive film formation on iron and film breakdown in a sodium hydroxide solution containing chloride ions. J. Electrochem. Soc. **132**(12), 2868–2871 (1985)
38. Brett, M.E., Parkin, K.M., Graham, M.J.: The passive film on iron: an investigation by electron back-scattering mössbauer spectroscopy. J. Electrochem. Soc. **133**(12), 2031–2035 (1986)
39. Haupt, S., Strehblow, H.H.: Corrosion, layer formation, and oxide reduction of passive iron in alkaline solution: a combined electrochemical and surface analytical study. Langmuir **3**(6), 873–885 (1987)
40. Bardwell, J.A., Macdougall, B., Graham, M.J.: Use of ^{18}O/SIMS and electrochemical techniques to study the reduction and breakdown of passive oxide films on iron. J. Electrochem. Soc. **135**(2), 413–418 (1988)
41. Boucherit, N., Delichere, P., Joiret, S., et al.: Passivity of iron and iron alloys studied by voltammetry and raman spectroscopy. Mater. Sci. Forum **44–45**, 51–62 (1989)
42. Larramona, G., Gutiérrez, C.: The passive film on iron at pH 1–14: a potential-modulated reflectance study. J. Electrochem. Soc. **136**(8), 2171–2178 (1989)
43. Gui, J., Devine, T.M.: In situ vibrational spectra of the passive film on iron in buffered borate solution. Corros. Sci. **32**(10), 1105–1124 (1991)
44. Gui, J., Devine, T.M.: The influence of sulfate ions on the surface enhanced Raman spectra of passive films formed on iron. Corros. Sci. **36**(3), 441–462 (1994)
45. Ryan, M.P., Newman, R.C., Thompson, G.E.: An STM study of the passive film formed on iron in borate buffer solution. J. Electrochem. Soc. **142**(10), L177–L179 (1995)

46. Davenport, A.J., Sansone, M.: High resolution in situ XANES investigation of the nature of the passive film on iron in a pH 8.4 borate buffer. J. Electrochem. Soc. **142**(3), 725–730 (1995)
47. Schroeder, V., Devine, T.M.: Surface enhanced raman spectroscopy study of the galvanostatic reduction of the passive film on iron. J. Electrochem. Soc. **146**(11), 4061–4070 (1999)
48. Joiret, S., Keddam, M., Nóvoa, X.R., et al.: Use of EIS, ring-disk electrode, EQCM and raman spectroscopy to study the film of oxides formed on iron in 1 M NaOH. Cement Concr. Compos. **24**(1), 7–15 (2002)
49. Sato, M., Kimura, M., Yamashita, M., et al.: Atomic-structure characterization of passive film of fe by grazing incidence X-ray scattering at SPring-8. In: MARCUS, P., MAURICE, V. Passivation of Metals and Semiconductors, and Properties of Thin Oxide Layers: A Selection of Papers from the 9th International Symposium, 95–100, Elsevier, Paris, France, 27 June–1 July 2005 (2006)
50. Deng, H., Nanjo, H., Qian, P., et al.: Potential dependence of surface crystal structure of iron passive films in borate buffer solution. Electrochim. Acta **52**(10), 4272–4277 (2007)
51. Harrington, S.P., Wang, F., Devine, T.M.: The structure and electronic properties of passive and prepassive films of iron in borate buffer. Electrochim. Acta **55**(13), 4092–4102 (2010)
52. Dormohammadi, H., Pang, Q., Murkute, P., et al.: Investigation of chloride-induced depassivation of iron in alkaline media by reactive force field molecular dynamics. npj. Mater. Degrad. **3**, 19 (2019)

Chapter 2
Determination Methods of Passivation

Abstract The definitions of passivatability and self-passivity, the typical potentiodynamic polarization curves of metals in active, passive, or pseudo-passive states, and some surface analysis methods for detecting the passive films are briefly introduced in this chapter.

2.1 Passivatability and Self-Passivity

For metallic passivation, Revie and Uhlig [1] gave the following two definitions (or situations):

(1) A metal is passive if it substantially resists corrosion in a given environment resulting from marked anodic polarization.
(2) A metal is passive if it substantially resists corrosion in a given environment despite a marked thermodynamic tendency to react.

Definition (1) means that the passivation potential of metal is higher than its corrosion potential. The one cannot be anodized, even if its corrosion current is very low, such as Fe in deoxidized aqueous solution, is excluded. Most transition metals and their alloys can be classified into this category, such as Cr, Ni, Co, W, Mo, Ti, stainless steel, etc. It includes the anodic polarization and the passivation by strong oxidizers, such as chromates [1].

Definition (2) means that the passivation is not due to anodic polarization, but due to the deposited diffusion barrier formed by the corrosion products on the metal surface. Such as, Pb forms $PbSO_4$ *deposit layer in* H_2SO_4*, Mg forms MgO hydrate in water, or Fe is inhibited in pickling solution. They are those metals which are characterized by low corrosion rate in the active potential range, i.e., very small anodic polarization* [1].

Kruger [2] *also divided the metallic passivation into the following two categories:*

(1) *A metal active in the electromotive force (emf) series is passive when its electrochemical behavior in a given environment becomes that of a metal noble in the emf series (low corrosion rate, noble potential).*

X. Lu, *Passivation and Corrosion of Black Rebar with Mill Scale*, Engineering Materials,
https://doi.org/10.1007/978-981-19-8102-9_2

(2) *A metal is passive while, still from the standpoint of thermodynamics at an active potential in a given environment, it exhibits a low corrosion rate (low corrosion rate, active potential). This type of passivity can be termed "practical passivity."*

Tomashov and Chernova [3] *divided the metals into five groups according to their standard hydrogen potential from low to high, i.e., from active to inactive. Please see Appendix A.*

Generally, the passivation caused by anodic polarization is called anodic oxidation or anodic passivation, and the passivation caused by the natural interaction between metal and the contact medium is called chemical passivation or self-passivation.

In this monograph, the anodic passivation characteristic is called "passivatability", and the chemical passivation characteristic is called "self-passivity".

2.2 Determination of Passivity

2.2.1 Potentiodynamic Polarization Method

Electrochemical methods are often used to determine the passivatability of metals, of which the most commonly used is the potentiodynamic polarization method.

Figure 2.1 is a schematic diagram of typical potentiodynamic polarization curves, including three anodic behaviors: active, passive, and pseudo-passive [4].

Here, E_c and i_c are self-corrosion (open circuit) potential and self-corrosion current (actually "current density", hereinafter referred to as current) respectively; E_{pp} and i_{cc} are critical passivation potential and critical current respectively; i_p is the passive

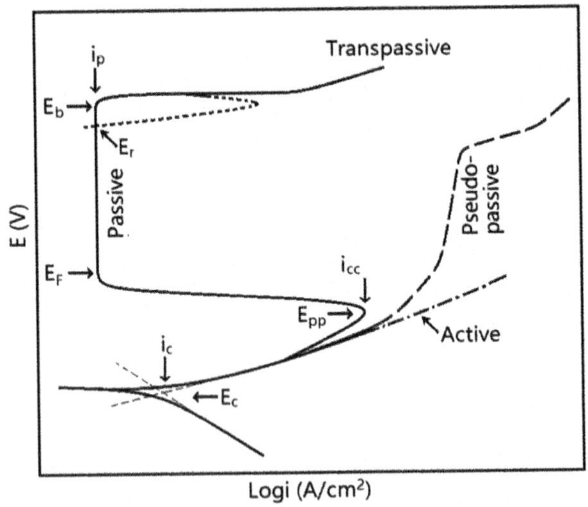

Fig. 2.1 Schematic diagram of typical potentiodynamic polarization curves, reprinted from [4] with permission of IOP Publishing, the "active" line is slightly extended

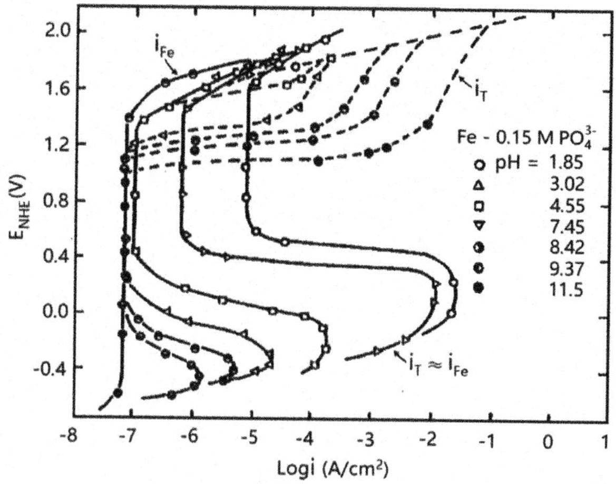

Fig. 2.2 Anodic polarization curves of iron in 0.15 M sodium phosphate solutions with different pH values, reprinted from [5] with permission of Springer Nature, the axes are rotated for a consistent context style

current; E_b is the pitting (passive film broken) potential; E_r is the repassivation potential, and E_F is the Flade (reactivation) potential.

As shown in Fig. 2.1, with the anodic potential increases:

(1) if the anodic current keeps increasing without decreasing, it means that the metallic electrode is in an active state, i.e., corrosion state;
(2) if a plateau of tiny current appears in an anodic potential range, it indicates that the metallic electrode is passivatable in this medium and may form a passive film, which needs to be further confirmed in combination with surface analysis;
(3) if the anodic current does not decrease significantly within an anodic potential range, but it increases very slowly or changes little, it may indicate that the metallic electrode has a pseudo-passivation behavior under this condition, which is usually related to a thick and porous deposit film-forming. Whether the protective layer is a passive film needs to be investigated together with other methods.

Figure 2.2 is the anodic polarization curves of Fe in 0.15 M Na_3PO_4 – NaOH solution with different pH values [5]. Where, i_T is the total current, and i_{Fe} is the anodic dissolution current.

It can be seen from Fig. 2.2 that when the pH value increases, the critical passivation potential and current both decrease remarkably, meanwhile, the passivation potential region becomes larger, and the passive current smaller, with the pitting potential decreases slightly. At pH = 11.5, the passivation potential region of Fe is about –0.4 V/NHE – 1.0 V/NHE,[1] and the passive current is about 0.7 μA/cm^2.

The anode current can reflect the rate of metallic dissolution (such as electrolysis or corrosion rate). If the reaction is known, the anode current can be converted into different forms of corrosion rate by Faraday's Law (see Eq. 2.1*)* [6]*:*

[1] NHE: Normal Hydrogen Electrode.

Table 2.1 *Classification of corrosion resistance of ordinary carbon steel, reprinted from* [6] *with permission of Springer Nature*

Corrosion rate		*Corrosion current*[a]	*Grade*
mpy	*mm/y*	$\mu A/cm^2$	
< 1	< 0.0254	< 2.2	*Outstanding*
1–5	0.0254–0.127	2.2–11	*Excellent*
5–20	0.127–0.508	11–43	*Good*
20–50	0.508–1.27	43–109	*Fair*
50–200	1.27–5.08	109–430	*Poor*
> 200	> 5.08	> 430	*Unacceptable*

[a] *Calculated according to* $Fe \rightarrow Fe^{2+} + 2e$, *the same as the following*

$$W = itm/nF \tag{2.1}$$

where, W is the weight loss of metal (g); i is the metallic electrolysis or corrosion current (A); t is time (s); m is the atomic weight of metal (g); n is the metallization valence; F is Faraday constant (96500C/mol).

If the corrosion current of Fe is 1 $\mu A/cm^2$, *and the reaction is* $Fe \rightarrow Fe^{2+}$, *i.e.,* $n = 2$, *then the corrosion rate calculated from Eq.* (2.1) *is about 12* $\mu m/y$.

Table 2.1 *lists the corrosion resistance classification for ordinary carbon steels and their corresponding corrosion currents* [6].

There are other different metallic corrosion grades also. For example, some think the corrosion rate less than 2 $\mu m/y$ *is negligible* [7], *between 2–5* $\mu m/y$ *is low, between 5–10* $\mu m/y$, *10–50* $\mu m/y$, *50–100* $\mu m/y$, *and* > *100* $\mu m/y$ *are medium, semi-high, high, and very high, respectively.*

Some gave the following five grades [8]*: < 0.05, 0.05–0.5, 0.5–1.5 and > 1.5 mm/y; or in ten grades* [9]*: < 0.001, 0.001–0.005, 0.005–0.01, 0.01–0.05, 0.05–0.1, 0.1–0.5, 0.5–1.0, 1.0–5.0, 5.0–10 and > 10 mm/y.*

People can choose different corrosion grades according to the different metals in different service media, or depending on the different corrosion stages or expected service lives.

For ordinary steel, when its corrosion rate is less than 1–2 $\mu m/y$, *i.e., the corrosion current* < *0.1–0.2* $\mu A\ cm^2$, *then it is generally considered to be in a completely passive state.*

Table 2.2 *lists the corrosion current and corresponding corrosion classification of reinforcement in concrete given by the European RILEM TC154 Committee* [10]. *It is considered that the corrosion current of reinforcement measured on-site by the recommended linear polarization resistance method is usually between 0.1–1* $\mu A/cm^2$, *greater than 1* $\mu A/cm^2$ *is rarely observed.*

It should be noted that the typical passive curve shown in Fig. 2.1 *is mostly the polarization behavior in acidic solution, that is, its passive current is usually significantly smaller than the critical current, which is sometimes several orders of*

Table 2.2 *Corrosion current and corrosion rating of reinforcement in concrete, reprinted from* [10] *with permission of Springer Nature*

Corrosion current ($\mu A/cm^2$)[a]	*Corrosion rate ($\mu m/y$)*	*Corrosion grade*
≤ 0.1	*≤ 1*	*Negligible*
0.1–0.5	*1–5*	*Low*
0.5–1	*5–10*	*Moderate*
> 1	*> 10*	*High*

[a] *By linear polarization resistance, the concrete resistance is also considered*

magnitude smaller. In the strong alkaline solutions, it is not always the same since metals may be self-passivated already.

In addition, the passive current can also reflect the quality of the passive film. The smaller the passive current, the denser the passive film. For those cases with typical passivation characteristics but with large "passive" currents, it should be careful to make sure if the metal is in a "true" or a "pseudo" passive state. It is necessary to analyze whether it has generated passive film or deposit film with help of other methods. It is not suitable to judge only by the potentiodynamic polarization curves.

2.2.2 Other Methods

Table 2.3 lists some commonly used passivation tests and surface analysis methods for metals [6, 11]. In practice, the passivation of metals is usually analyzed by two or more methods, such as potentiodynamic polarization or natural immersion plus Ellipsometry, Raman, XPS, Auger, or Mössbauer spectroscopies.

Except for the potentiodynamic polarization, the galvanostatic anodic polarization and potentiostatic anodic polarization methods among the electrochemical methods can be used to investigate the passivation, too.

Figure 2.3 *demonstrates a galvanostatic anodic polarization test* [12] *which was previously used to test the corrosivity of chemical admixtures to reinforcement, it was once used to determine the passivation behavior of rebar in the past, i.e., applying an appropriate anodic polarization current (such as 50 $\mu A/cm^2$) to the metallic electrode to see:*

(1) *if its potential moves forward rapidly and reaches a stable value in a short time (such as 1–5 min), and does not decrease after an appropriate time (such as 30 min), as the curve (1) shown in Fig.* 2.3, *then the metal is passivatable in the measured medium and can produce a dense and passive film;*
(2) *if the potential of the metallic electrode is significantly positive at first and then gradually decreases, as the curve (2) shown in Fig.* 2.3, *then the metal may be passivatable, but the passive film may be partially damaged;*
(3) *the curve (3) in Fig.* 2.3 *indicates that the metal may be between passive and active states; even if a passive film can be formed, it may be unstable or can be*

Table 2.3 Metallic passivation tests and analysis methods, reprinted from [6] with permission of Springer Nature

Type	Methods	Index
Electrochemical	Potentiodynamic polarization	E_p, i_p
	Galvanostatic anodic polarization	E-t
	Potentiostatic anodic polarization	i-t
	Galvanostatic reduction	Film thickness, electric quantity
	Cyclic voltammetry (CV)	Peaks, electric quantity
	Rotating ring disc electrodes	Reaction rates
Optical	Spectroscopic Ellipsometry	Amplitude, phase, refractive index
	Infrared spectroscopy (IR)	Group analysis
	Ultraviolet spectroscopy (UV)	Group analysis
Electron optical	Auger electron spectroscopy (AES)	Element and valence state
	Ion scattering spectroscopy (ISS)	Isotope, impurity
	Mössbauer spectroscopy (Mössbauer)	Elements and compounds
	Raman spectroscopy (Raman)	Elements and compounds
	Rutherford backscattering (RB)	Element depth distribution
	Secondary Ion Mass Spectrometry (SIMS)	Elements
	Scanning force microscopy (SFM)	Atomic arrangement, morphology
	Scanning tunneling microscope (STM)	Atomic arrangement, morphology
	Transmission electron microscope (TEM)	Crystal constants and orientations
	X-ray absorption near-edge structure (XANES)	Valence state
	X-ray photoelectron spectroscopy (XPS)	Binding energy
	X-ray diffraction (XRD)	Composition, crystal constants
	X-ray absorption spectroscopy (XAS)	Element, valence state

seriously damaged; at this time, its passivity is not easy to judge only by this method;

(4) *if the potential of the metallic electrode does not increase or decrease continuously, as the curve (4) shown in Fig. 2.3, then the metal cannot produce a passive film under such a condition and keep in the active state.*

Since the methods using a single fixed current (or potential) have some shortcomings obviously, such as the applied current (or potential) is not always making the metal fall into the passive region in the contact medium, and when the applied current is too large or too small, there may be over-passivated or under-passivated. At present, those methods are seldomly used or abandoned.

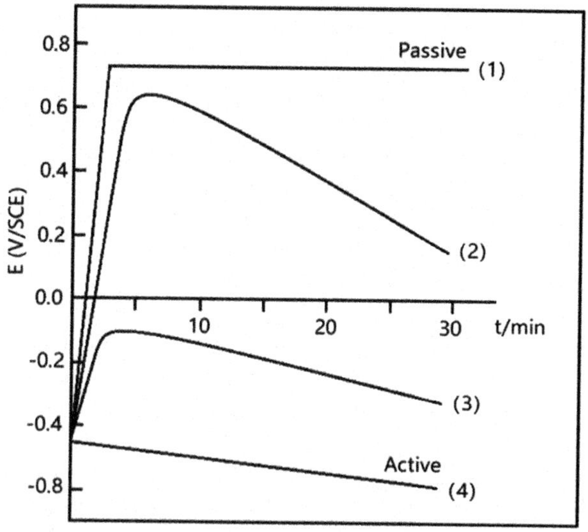

Fig. 2.3 *Galvanostatic anodic polarization* [12][2]

Galvanostatic reduction was widely used in the early decades of last century, but sometimes it provides wrong information about the "multilayer" characteristics of the passive film. First, the passive film may undergo phase transformation during the test, then the applied current is not all used on electrolysis; Second, even a "single" layer of passive film may have two-stage polarization steps, so the presence of two reduction steps does not always mean the passive film has a bilayer structure.

Like the potentiodynamic polarization, the cyclic voltammetry can also have a low current density plateau in a certain potential region when the metal is passivatable in the tested medium (see Chap. 6). However, the cyclic voltammetry is mostly used for the study of metallic electrode reactions and their reversibility.

The rotating disk electrode is mainly used to study the electrode process, especially the one with adsorption or diffusion reaction. By analyzing the specific electrode reaction parameters, the passivation phenomenon can also be figured out.

The optical and electron-optical methods are commonly employed to determine whether a passive film is formed on the metal surface. Among them, ellipsometry is a very useful method. It can monitor the formation process of a passive film in solution at real-time and in-situ. Unfortunately, it cannot directly provide the chemical composition about the passive film. When the passive film is a single layer and the refractive index is fixed, then the thickness of the passive film can be calculated accurately, otherwise, it cannot. Because the ellipsometry is very sensitive to the change of material surface, it is often used to observe the initial on-site growth of the passive film.

Most of the energy spectrum analysis methods usually require high vacuum and focused high-energy electron beams. The vacuum and the electron beam irradiation may cause the dehydration or decomposition of the films sometimes, resulting in

[2] SCE: Saturated Calomel Electrode.

inconsistency between the observed results and the original sates, which often leads to controversy and divergence. In addition, the surface analysis methods often require the surface of the specimen to be mirror-smooth, which limits their application to the specimens with rough surfaces, such as the rebar with mill scale.

Those optical analysis techniques which are not affected by the irradiation, such as infrared spectroscopy, are not easy to exclude the influence of water molecules or hydroxyl groups adsorbed on the metal surface. Therefore, it should be more careful in analyzing the results at not dry enough conditions, otherwise, it is easy to draw conclusions wrong.

When the medium contains organic substances, such as the corrosion inhibitor or other chemical admixtures, the open-circuit potentials of metals are usually decreased and the electrode processes might be changed. So, it is not suitable to discuss the passivity of the metal by the oxide film forming theory under such conditions. It may be comprehensively analyzed based on electrochemical tests in combination with the adsorption isotherm plus the infrared, ultraviolet, mass spectrometry, nuclear magnetic resonance methods, or other results obtained by different technologies. Since it is beyond the scope of this book, it will not be described here.

2.3 Summary

The methods how to determine the passivity of metals are briefly introduced in this chapter. Among them the potentiodynamic polarization is mainly described. The typical anodic polarization curves of metals in active, passive, and pseudo-passive states are given. To make sure if there is a passive film formed or not, at least two or more different methods need to be employed generally.

References

1. Revie, R.W., Uhlig, H.H.: Corrosion and Corrosion Control: An Introduction to Corrosion Science and Engineering, 4th edn., p. 84. John Wiley & Sons Inc., New Jersey (2008)
2. Kruger, J.: Passivity in corrosion: Fundamentals, testing, and protection. In: ASM Handbook, vol. 13A, pp. 61−67. ASM International, Ohio (2003)
3. Tomashov, N.D., Chernova, G.P.: Passivity and Protection of Metals Against Corrosion. Plenum Press, New York (1967)
4. Viramontes-gamboa, G., Rivera-vasquez, B.F., Dixon, D.G.: The active-passive behavior of chalcopyrite. J. Electrochem. Soc. **154**(6), C299–C311 (2007)
5. Sato, N., Okamoto, G.: Electrochemical Passivation of Metals. In: Bockris, J.O'.M., Conway, B.E., Yeager, E., et al. (eds.)Electrochemical Materials Science. Comprehensive Treatise of Electrochemistry, vol. 4, pp. 193−245. Springer, Boston (1981)
6. McCafferty, E.: Introduction to Corrosion Science, vol. 13–32, pp. 209–262. Springer, New York (2010)
7. Bertolini, L., Elsener, B., Pedeferri, P., et al.: Corrosion of Steel in Concrete: Prevention, Diagnosis, Repair. WILEY-VCH, Weinheim (2004)

8. Zuo, J., Zuo, Y.: Corrosion Data and Material Selection Manual. Chemical Industry Press, Beijing (2008)
9. Zhu, R., Gu, G., Yang, W., et al.: Metallic Corrosion Protection Manual. Shanghai Science and Technology Press, Shanghai (1989)
10. Andrade, C., Alonso, C.: Test methods for on-site corrosion rate measurement of steel reinforcement in concrete by means of the polarization resistance method. RILEM TC 154-EMC recommendations. Mater. Struct./Matériaux Constr. **37**, 623–643 (2004)
11. Strehblow, H.H.: Passivity of metals studied by surface analytical methods, a review. Electrochim Acta **212**(10), 630–648 (2016)
12. GB 8076−1997.: Concrete Admixtures. Appendix B Rapid test method for reinforcement corrosion (Fresh mortar test)

Chapter 3
Passivation of Rebars Without Mill Scale

Abstract The passivation behavior of the black rebar without mill scale is briefly introduced. The passive film on the descaled rebar generated by anodic polarization is mainly composed of γ-Fe_2O_3 and Fe_3O_4, and the one produced by natural immersion is mainly Fe_3O_4 or Fe_3O_4 plus some γ-Fe_2O_3 mixed in the outer layer.

3.1 Potentiodynamic Polarization of Rebars Without Mill Scale

From this chapter, it will be described according to some our experimental results. Here the rebar without mill scale refers to that of whose mill scale removed by mechanical brushing or pickling, and the one of polished cross-section.

Black rebars are generally divided into hot-rolled, cold-rolled, and cold-drawn rebars. The hot-rolled rebars can also be divided into hot-rolled plain bar (HPB), Hot-rolled ribbed bar (HRB), retained heat treatment ribbed bar (RRB), and hot-rolled bar of fine grains (HRBF).

The type of rebar is usually labelled by its rolling method and yield strength. For example, HPB235 is a hot-rolled plain rebar with a yield strength of 235 MPa, while HRB335 and HRB400 are the hot-rolled ribbed rebars with a yield strength of 335 and 400 MPa, respectively. Traditionally, the reinforcement is also recorded as grade I–IV from low to high yield strength. For example, the three kinds of rebars mentioned above are also called grade I, II, and III reinforcements, respectively. However, these terms may change in the future, because the steel bars with yield strength less than 235 and 335 MPa have been canceled in the versions of Steel for reinforced concrete Part 2: Hot rolled ribbed steel bars GB/T 1499.2−2007 and 2018, respectively, and only three yield strength grades of 400, 500 and 600 MPa have been retained in the 2018 version.

The most commonly used HPB235, HRB335, and HRB400 reinforcements were employed in our studies, and the corresponding test results in the following parts are extracted from [1–3], *respectively.*

X. Lu, *Passivation and Corrosion of Black Rebar with Mill Scale*, Engineering Materials,
https://doi.org/10.1007/978-981-19-8102-9_3

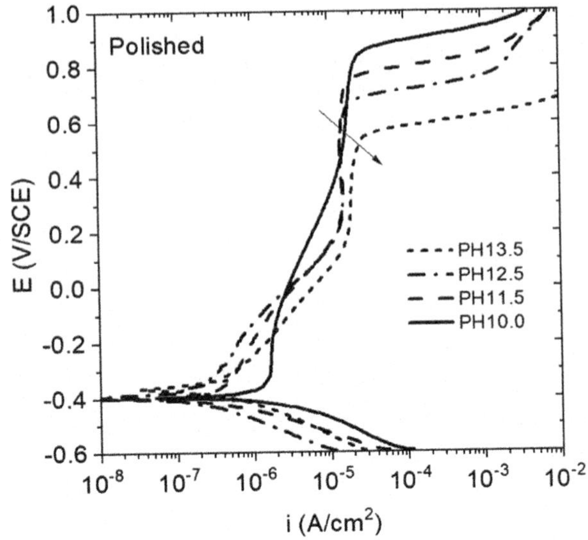

Fig. 3.1 Potentiodynamic polarization curves of polished HPB235 rebar

Figure 3.1 is the potentiodynamic polarization curves of polished Φ8mm HPB235 rebar in simulated concrete pore solution with different pH values. In order to eliminate the influence of oxide on the rebar surface formed during polishing, potential scanning from cathode to anode was carried out [1].

The simulated concrete pore solution is 0.6 mol/l KOH + 0.2 mol/l NaOH + saturated $Ca(OH)_2$ solution, and the pH value is adjusted by saturated $NaHCO_3$ solution. The pH value is measured by a pH-2s type meter, and the potentiodynamic polarization by Solartron 1287 electrochemical station with a scan rate of 1 mV/S.

The polished HPB235 rebar shows a certain passivation at pH = 11.5 and 12.5 in the passive potential regions of 0.3–0.7 V and 0.3–0.6 V, respectively, and the passive currents both are around 15 μA/cm^2. At pH = 10.0 and 13.5, it demontrates the "pseudo-passivation" behavior, and the "passive" currents in the potential regions of 0.5–0.8 V and 0.2–0.5 V are about 18 μA/cm^2 and 20 μA/cm^2, respectively.

Figure 3.2 is the potentiodynamic polarization curves of polished and pickled Φ12 mm HRB335 rebars in saturated $Ca(OH)_2$ solution obtained by CS350 electrochemical workstation with a scan rate of 0.2 mV/s. Before dynamic potential scanning, the two kinds rebar samples were polarized at −1.5 V for 5 min to remove the surface oxide generated in the air [2].

The descaled HRB335 rebar has an obvious passivation behavior in saturated $Ca(OH)_2$ solution. The passive potential range of polished rebar is about −0.5 – 0.5 V, and the passive current is around 6 μA/cm^2. The passive potential range of the pickled one is about −0.2 – 0.5 V, and the passive current is about 12 μA/cm^2. It indicates that the surface state of rebar can significantly affect the passivation characteristics, and the rough surface usually needs a large current to passivate.

Figure 3.3 is the multiple dynamically scanned curves of the mechanically descaled Φ12 mm HRB400 rebar in saturated $Ca(OH)_2$ solution with a scan rate

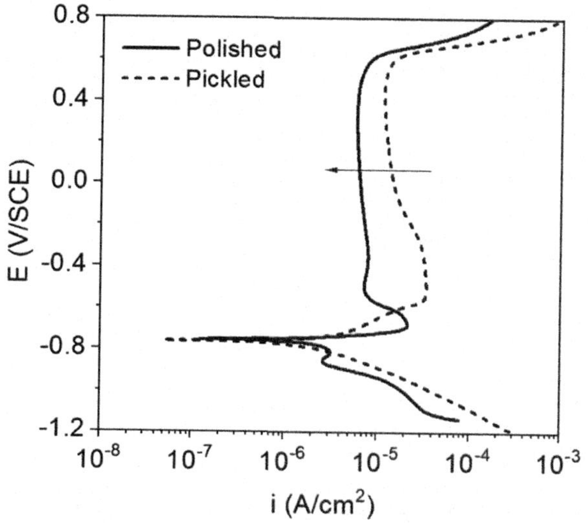

Fig. 3.2 Potentiodynamic polarization curves of HRB335 rebar

of 1 mV/s [3]. The first scan starts after the sample immersed in solution for 15 min, and the interval between the two scans is 2 h.

The descaled HRB400 rebar also has an obvious passivation behavior in saturated $Ca(OH)_2$ solution. The passive potential region is about $-0.2 - 0.6$ V with a passive current around 35 $\mu A/cm^2$. So, the rougher the rebar surface, the greater the passive current, and less dense the passive film.

It can be seen that after the first potentiodynamic polarization in Fig. 3.3, the descaled HRB 400 rebar is passivated, its self-corrosion potential increases from

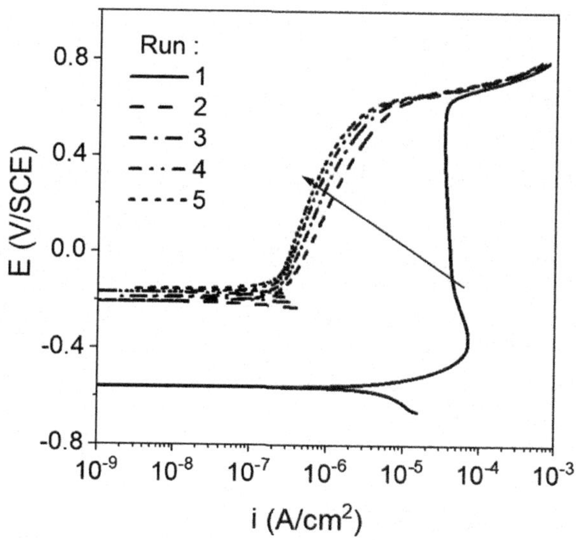

Fig. 3.3 Potentiodynamic polarization curves of HRB400 descaled rebar

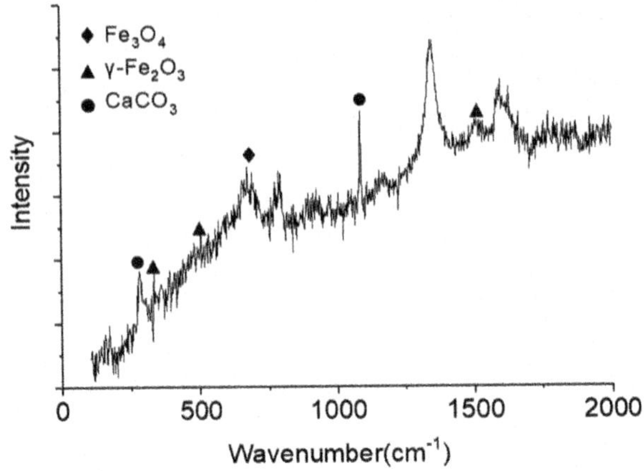

Fig. 3.4 Surface Raman spectrum of potentiostatically polarized HRB335 descaled rebar

−0.56 to −0.20 V with the anodic polarization curve moves significantly to the upper left, and the anodic polarization current corresponding to −0.1 – 0.5 V of 0.3–5 μA/cm^2 gradually decreases to 0.2–2 μA/cm^2 from the second to the fifth scan, while the polarization curve shows the "pseudo-passivation" characteristics.

From Figs. 3.1, 3.2 and 3.3, it can be seen that the anodic polarization behavior of different descaled rebars is different. The results shown in Fig. 3.3 can also remind us that it should be careful when we determine the passivity of metals by the potentiodynamic polarization method, do not mechanically judge it only by the shape of the anodic polarization curve. Before giving a conclusion, it's better to consider the surface characteristics, the magnitude of passive current, and whether there is a passive film already on the surface as well as other factors, then to make a comprehensive analysis in combination with other methods.

3.2 Passive Films of Rebars Without Mill Scale

3.2.1 *Produced by Anodic Polarization*[1]

Raman spectroscopy is chosen to inspect the composition change of rebar surface before and after tests.

Figure 3.4 is the Raman spectrum of the descaled HRB335 rebar after intermittent potentiostatic passivation for 140 days in saturated $Ca(OH)_2$ solution at 0.35 V. The potentiostatic passivation duration is 2 h every 10 days.

Except the $CaCO_3$ produced by the carbonization of solution, the passive film obtained by anodic passivation on the descaled HRB335 rebar is composed of γ-Fe_2O_3 and Fe_3O_4.

[1] The passive film of HPB235 steel has been extensively studied, so Raman test was not repeated.

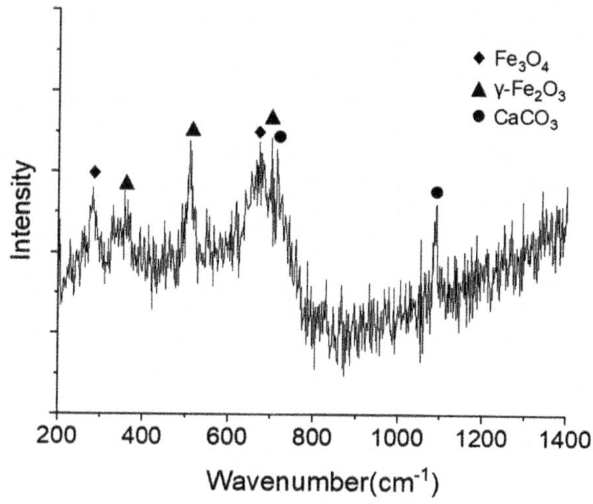

Fig. 3.5 Surface Raman spectrum of potentiostatically polarized HRB400 descaled rebar

Figure 3.5 is the Raman spectrum of the descaled HRB400 rebar after continuous passivation at 0.3 V for 60 days in saturated $Ca(OH)_2$ solution. After eliminating the $CaCO_3$ generated by the solution carbonization, the passive film of the descaled HRB400 rebar is also composed of γ-Fe_2O_3 and Fe_3O_4.

3.2.2 Produced by Natural Immersion

Figure 3.6 is the Raman spectrum of the descaled HRB335 rebar after continuous natural immersion in saturated $Ca(OH)_2$ solution for 140 days. Different from the potentiostatic polarization results in Fig. 3.4, the passive film generated on the descaled HRB335 rebar after a long-term immersion in alkaline solution is Fe_3O_4.

Figures 3.7 and 3.8 are Raman spectra of the descaled HRB400 rebars after natural immersion in saturated $Ca(OH)_2$ solution for 7 days and 360 days, respectively.

When the natural immersion time is short, the passive film of the descaled HRB400 rebar is Fe_3O_4; but when the natural immersion time is long, the passive film is composed of γ-Fe_2O_3 and Fe_3O_4. It should be noted that there is only Fe_3O_4 at some sites (as site B), that means, γ-Fe_2O_3 and Fe_3O_4 coexist on the surface, not the γ-Fe_2O_3 covers Fe_3O_4 completely.

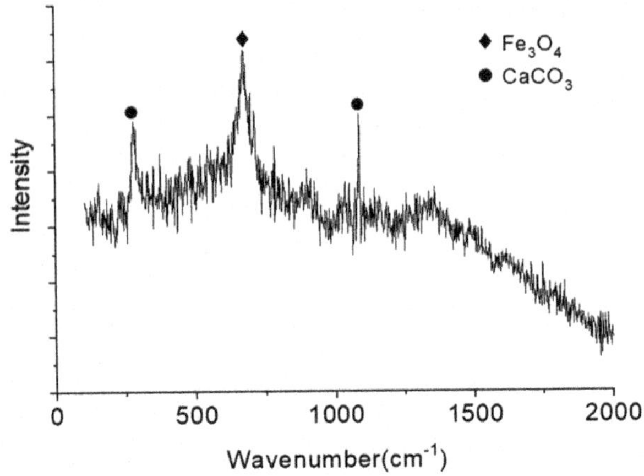

Fig. 3.6 Surface Raman spectrum of HRB335 descaled rebar after natural immersion of 140d

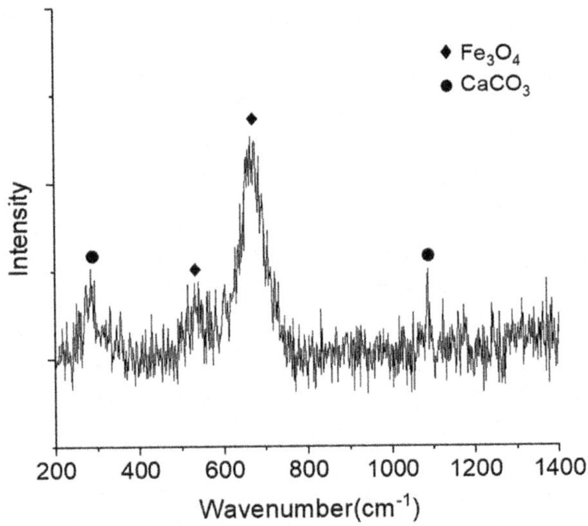

Fig. 3.7 Surface Raman spectrum of HRB400 descaled rebar after natural immersion 7d

3.3 Passivation Characteristics of Descaled Rebars

It can be seen from the above experimental results that the descaled rebar has passivatability and self-passivity under strongly alkaline conditions. When the rebar is anodically polarized, the passive film is mainly composed of $\gamma\text{-}Fe_2O_3$ and together with Fe_3O_4. When the rebar is chemically passivated, the passive film is mainly composed of Fe_3O_4, and together with $\gamma\text{-}Fe_2O_3$. This is almost consistent with the previous publications, as shown in Table 3.1.

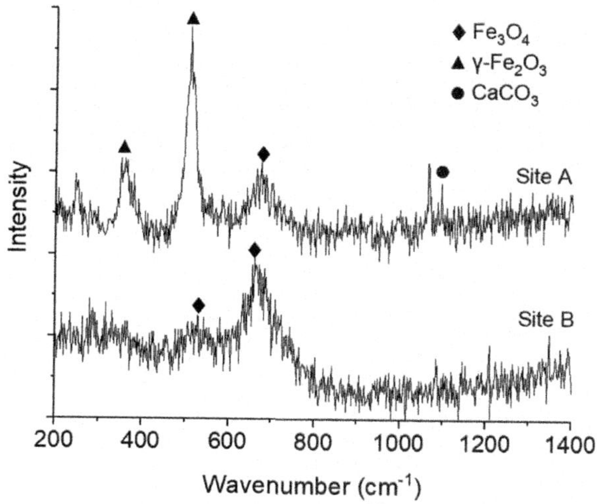

Fig. 3.8 Surface Raman spectrum of HRB400 descaled rebar after natural immersion 360d

According to Table 3.1, the descaled black rebar, similar to the pure Fe, can be passivated in strong alkaline aqueous solutions. The generated passive film mostly shows the n-type semiconductor transfer characteristics, and has a bilayer structure. The inner layer is a Fe^{2+} oxide or Fe_3O_4, and the outer layer is a Fe^{3+} oxide or hydroxide, such as α-Fe_2O_3, γ-Fe_2O_3, α-FeOOH, or γ-FeOOH. The specific composition of passive film varies with the solution composition, pH value, and passivation process. In general, under strong oxidation conditions, the α−phase is obtained, and under secondary oxidation conditions, the γ-phase is obtained.

3.4 Summary

The descaled black rebar can produce passive film on the surface, which is composed of Fe_3O_4 or γ-Fe_2O_3, or their mix, in simulated concrete pore solutions. The other passive films on the descaled rebars, such as α-Fe_2O_3, α-FeOOH, and γ-FeOOH reported in the publications are not found in the present work.

Table 3.1 Passive films formed on descaled reinforcements under different conditions

Test method	Solution	pH	Surface roughness	Passivating condition	Compositions	Semi-conductor	Thickness (nm)	Ref.
Photoelectrochemical	KOH, NaOH, $Ca(OH)_2$	12.5–13.5	600#	–	γ-FeOOH	n	–	Chu et al. [4]
Photoelectrochemical	H_3BO_3, $Na_2B_4O_7$	8.5	–	0.4 V	γ-Fe_2O_3	n	2.9	Kim et al. [5]
				0.6 V 0.8 V	γ-FeOOH γ-Fe_2O_3	n	–	
XPS, AES	KOH, NaOH	13.0	1 μm	−0.8 – 0.3 V	$Fe^{3+}O$ $Fe^{2+}O$	–	–	Freire et al. [6]
XPS	KOH, NaOH, $Ca(OH)_2$	12.5 13.3	1 μm	Immersion, 15d	FeOOH, $Fe^{3+}O$ $Fe^{2+}O$	–	3 –13	Ghods et al. [7, 8]
Mott-Schottky	$Ca(OH)_2$	12.5	600#	–	–	n	–	Zhang et al. [9]
Mott-Schottky	KOH, NaOH, $Ca(OH)_2$		1200#	Immersion, 30 min	–	n	–	Liu et al. [10]
Mott-Schottky	H_3BO_3, $Na_2B_4O_7$	8.4	600#	–	–	n		Guo [11]
Raman	NaOH	14	1200#	−0.90 V	Fe_3O_4	–	–	Joiret et al. [12]
				−0.53 V	γ-FeOOH, Fe_3O_4			
				0.5 V	Fe_2O_3/FeOOH, Fe_3O_4			
Raman Mott-Schottky	H_3BO_3, $Na_2B_4O_7$	8.4	2400#	−0.56 – 0.444 V	$Fe^{3+}O$ γ-Fe_2O_3/Fe_3O_4	n	–	Harrington[13, 14]

(continued)

Table 3.1 (continued)

Test method	Solution	pH	Surface roughness	Passivating condition	Compositions	Semi-conductor	Thickness (nm)	Ref.
Raman	KOH	14	2 μm	−0.6 – 0.2 V	γ-FeOOH/α-FeOOH Fe_3O_4	–	–	Johnston[15]
Raman XPS	$Na_2B_4O_7$, NaOH	10.6	1000#	−0.6 V polarized 7d	Fe_3O_4	–	–	Xu et al. [16]
				−0.3 V	γ−Fe_2O_3, Fe_3O_4			
				0.2 V	γ-FeOOH/γ-Fe_2O_3 Fe_3O_4			
EELS	KOH, NaOH, $Ca(OH)_2$	13.3	1 μm	Immersion 15d	Fe_3O_4, $Fe^{2+}O$	–	–	Gunay et al. [17]
	$Ca(OH)_2$	12.5			α-Fe_2O_3/Fe_3O_4, $Fe^{2+}O$			
EIS	KOH, $Ca(OH)_2$	13.2	–	Immersion	Fe_3O_4	–	–	Sánchez et al. [18]

References

1. Lin, W.: Influences on rebar corrosion by rebar surface states, pH of concrete pore solution and [Cl^-] in concrete. Tsinghua University. Master thesis (2002). In Chinese
2. Li, Y.J.: Chemical stability of rebar with mill scale in simulated concrete pore solutions. Tsinghua University. Master thesis (2016). In Chinese
3. Wang, X.: Redox of rebar with mill scale in simulated concrete pore solutions. Tsinghua University. Master thesis (2019). In Chinese
4. Chu, W., Shi, Y., Wei, B., et al.: The photoelectrochemical studies of passive films on rebar electrode in simulated cement pore solution. Electrochemistry **03**, 291–297 (1995). in Chinese
5. Kim, J.S., Cho, E.A., Kwon, H.S.: Photoelectrochemical study on the passive film on fe. Corros. Sci. **43**(8), 1403–1415 (2001)
6. Freire, L., Nóvoa, X.R., Montemor, M.F., et al.: Study of passive films formed on mild steel in alkaline media by the application of anodic potentials. Mater. Chem. Phys. **114**(2–3), 962–972 (2009)
7. Ghods, P., Isgor, O.B., Brown, J.R., et al.: XPS depth profiling study on the passive oxide film of carbon steel in saturated calcium hydroxide solution and the effect of chloride on the film properties. Appl. Surf. Sci. **257**(10), 4669–4677 (2011)
8. Ghods, P., Isgor, O.B., Bensebaa, F., et al.: Angle-resolved XPS study of carbon steel passivity and chloride-induced depassivation in simulated concrete pore solution. Corros. Sci. **58**, 159–167 (2012)
9. Zhang, Y., Shi, M., Chen, Z.: Mott-Schottky investigation of passive film of rebar in simulated concrete pore solution. Mater. Mech. Eng. **07**, 7–10 (2006). in Chinese
10. Liu, Y., Du, R., Lin, C.: Study on semiconductor characteristics of rebar passive film in simulated concrete pore solution. In: National Conference on Corrosion Electrochemistry and Test Methods Xiamen, pp. 106–110 (2006). In Chinese
11. Guo, H.X., Lu, B.T., Luo, J.L.: Study on passivation and erosion-enhanced corrosion resistance by mott-schottky analysis. Electrochim. Acta **52**(3), 1108–1116 (2006)
12. Joiret, S., Keddam, M., Nóvoa, X.R., et al.: Use of EIS, ring-disk electrode, EQCM and raman spectroscopy to study the film of oxides formed on iron in 1 M NaOH. Cement Concr. Compos. **24**(1), 7–15 (2002)
13. Harrington, S.P., Devine, T.M.: Relation between the semiconducting properties of a passive film and reduction reaction rates. J. Electrochem. Soc. **156**(4), C154–C159 (2009)
14. Harrington, S.P., Wang, F., Devine, T.M.: The structure and electronic properties of passive and prepassive films of iron in borate buffer. Electrochim. Acta **55**(13), 4092–4102 (2010)
15. Johnston, C.: In situ laser raman microprobe spectroscopy of corroding iron electrode surfaces. Vib. Spectrosc. **1**(1), 87–96 (1990)
16. Xu, W., Daub, K., Zhang, X., et al.: Oxide formation and conversion on carbon steel in mildly basic solutions. Electrochim. Acta **54**(24), 5727–5738 (2009)
17. Gunay, H.B., Ghods, P., Isgor, O.B., et al.: Characterization of atomic structure of oxide films on carbon steel in simulated concrete pore solutions using EELS. Appl. Surf. Sci. **274**, 195–202 (2013)
18. Sánchez, M., Gregori, J., Alonso, C., et al.: Electrochemical impedance spectroscopy for studying passive layers on steel rebars immersed in alkaline solutions simulating concrete pores. Electrochim. Acta **52**(27), 7634–7641 (2007)

Chapter 4
Mill Scale of Hot-Rolled Rebars

Abstract The formation, composition, and microstructure of mill scale are briefly introduced. The mill scale formed at a relatively higher temperature with a rapid cooling is usually composed of FeO, Fe_3O_4, and Fe_2O_3 from inside to outside, forming a trilayer structure. Nowadays it is mainly composed of Fe_3O_4 obtained by the controlled rolling and cooling process for high strength rebars.

4.1 High-Temperature Oxidation of Iron

The mill scale of rebar is not a passive film. It is a thick oxide layer formed by the oxidation of rebar during high-temperature rolling. Before discussing the passivation of hot-rolled rebars, it is necessary to give a description about the mill scale in details first.

Figure 4.1 is the phase diagram of Fe–O system [1]. Fe reacts with oxygen at high temperatures producing FeO, Fe_3O_4, and Fe_2O_3, in which the oxygen content increases.

Figure 4.2 illustrates the section structure of oxide scale formed on Fe at different temperatures under static oxidations. Above 570 °C, the oxide scale is composed of FeO, Fe_3O_4, and Fe_2O_3 from inside to outside, forming a trilayer scale. Below 570 °C, FeO can transform into Fe and Fe_3O_4, so the oxide scale may be composed of Fe_3O_4 and Fe_2O_3 in a bilayer structure [2, 3].

Figure 4.3 shows the oxidation mechanism of Fe above 570 °C [3]. Because the migration rate of Fe^{2+} ions and electrons in FeO is much higher than that of them in Fe_3O_4 and Fe_2O_3, the relative thickness ratio of FeO: Fe_3O_4: Fe_2O_3 oxide scale generated by the static oxidation at 1000 °C is about 95:4:1 [2]. However, after repeated hot rolling (non-static oxidation), it is not easy to find the Fe_2O_3 oxide layer on the steel surface, only a small amount of Fe_2O_3 inclusions or residues are occasionally visible.

The eutectoid reaction of FeO below 570 °C is often used to convert the loose FeO layer generated at high temperature into a relatively dense and stable Fe_3O_4 scale through an appropriate cooling process to improve the atmospheric corrosion resistance of steel [2, 3].

X. Lu, *Passivation and Corrosion of Black Rebar with Mill Scale*, Engineering Materials,
https://doi.org/10.1007/978-981-19-8102-9_4

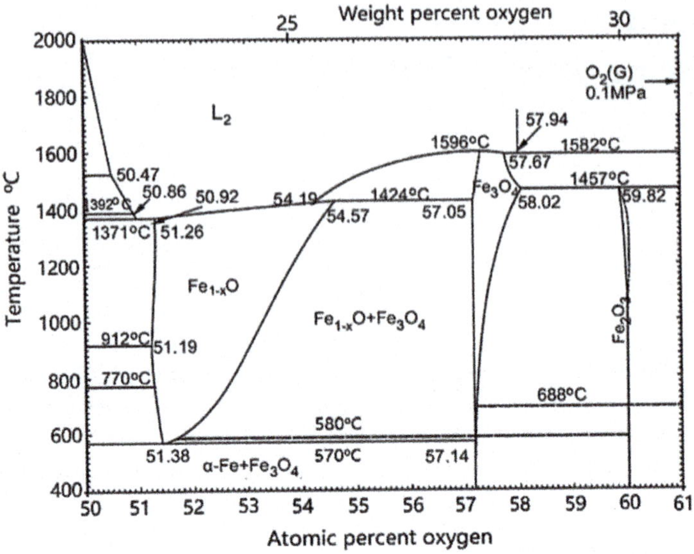

Fig. 4.1 Fe−O phase diagram, reprinted from [1] with permission of Springer Nature

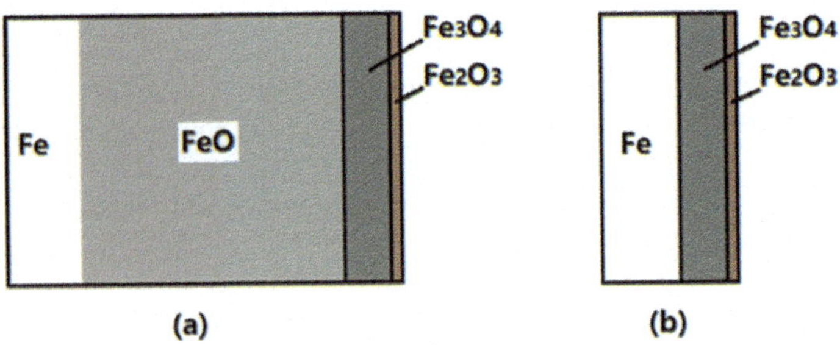

Fig. 4.2 Schematic diagram of oxide scale section structure generated by Fe in oxygen, (a) above 570 °C and (b) below 570 °C

Considering the valence of iron, Fe_2O_3 is the most stable oxide at room temperature. Fe_3O_4 is not completely chemically stable. In fact, when Fe_3O_4 is heated above 200 °C, it will become γ-Fe_2O_3, and if heated above 300 °C, it will become α-Fe_2O_3, i.e., Fe_3O_4, is more easily oxidized to γ-Fe_2O_3, owing to their similar crystal structure [2, 3]. So α-Fe_2O_3 is more stable than γ-Fe_2O_3. In addition, Fe_3O_4 and γ-Fe_2O_3 are magnetic, while α-Fe_2O_3 is non-magnetic.

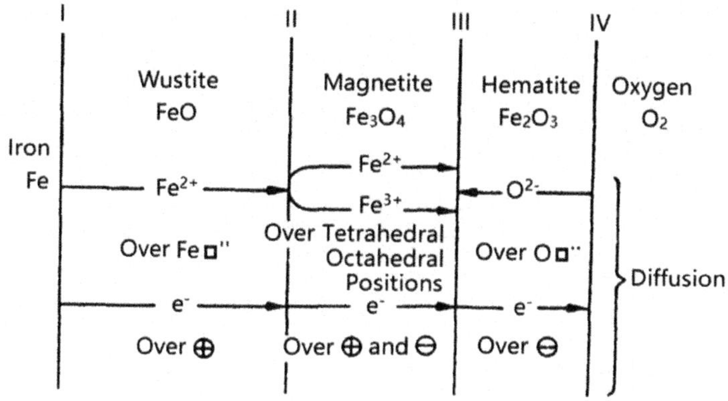

Fig. 4.3 Formation mechanisms of oxide scale on iron surface above 570 °C, reprinted from [3] with permission of Springer Nature

4.2 Formation of Mill Scale

4.2.1 Oxidation Process

Figure 4.4 is a schematic diagram of the high-temperature static oxidation process of metals [3]. Oxygen atoms are first chemically adsorbed at the active points on the metal surface. With the increase of adsorption capacity, local oxide nucleation occurs. When the adsorption is saturated, a single-layer oxide film is formed. With the further oxidation and atomic rearrangement of metal atoms, the metallic oxide nuclei are formed by local epitaxy and begin to grow laterally and longitudinally. In this way, the three-dimensional oxide scales are gradually formed.

The thickness growth of oxide scale follows different relationships at different oxidation stages. At the initial stage it generally follows the linear law, but in the middle or the later stage it may follow a logarithmic or parabolic law [2, 3]. The growth of oxide scale is usually controlled by the inward diffusion of oxygen atoms

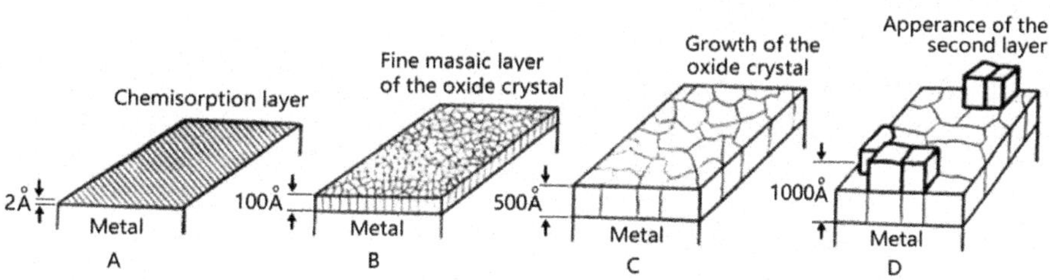

Fig. 4.4 High-temperature oxidation process of metals, reprinted from [3] with permission of Springer Nature

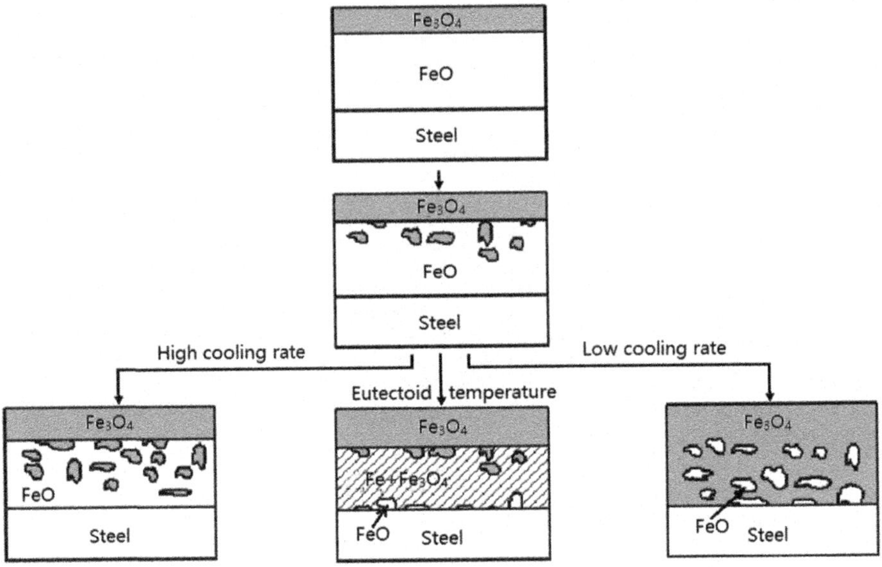

Fig. 4.5 Effect of cooling temperature rate on mill scale, reprinted from [4] with permission of Springer Nature

and the outward diffusion of metallic atoms. Generally, it grows relatively fast at the defects such as grain boundaries [2, 3].

According to the Fe–O phase diagram in Fig. 4.1, it can be known that at 570 °C, the eutectoid transformation will occur. Therefore, the actual oxide scale composition obtained at different cooling rates will be different. Figure 4.5 indicates the cooling effects on the composition of oxide scales [4]. The slower the cooling rate, the more thorough the FeO decomposition. Under appropriate conditions, it can form a sole Fe_3O_4 scale [5].

4.2.2 *Rolling Defects*

The oxidation time of billet in heating furnace is relatively long. The thick oxide scale generated in heating furnace is called primary oxide scale, which needs to be removed with high-pressure water before rolling. During the roughing and finishing rolling [6, 7], see Fig. 4.6, the oxide scale generated on the steel surface is called secondary and tertiary oxide scale, respectively, which is mostly called mill scale for short. Indeed, the final mill scale on the steel products is obtained by cooling after rolling, which is usually subjected to the phase transformation mentioned above.

Table 4.1 lists five typical mill scales on steels together with their defect formation mechanisms [8]. In the process of rough rolling and finish rolling, the residual oxide scale and secondary oxide scale that are not removed completely may be pressed into

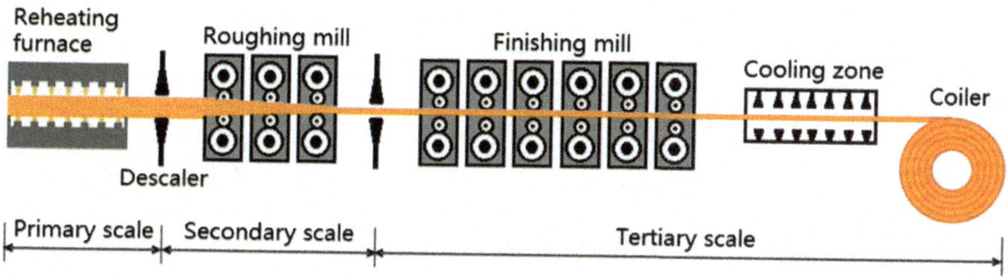

Fig. 4.6 Different oxide scales formed during the hot-milling process

the substrate surface. At the cracks of oxide scale, the substrate will be extruded out and get oxidized. If the oxide scale is debonding from the substrate and blisters born, a thick mill scale and cracks under it will appear thereafter rolling, see Fig. 4.7. When the roller surface is damaged, the corresponding marks will be inevitably printed on the product surface [5, 8–12].

Because the rolling time of each pass is short and the reduction is not exactly the same, the oxide scale generated during rolling is quite different from that at static oxidation on the thickness, composition, and the micro defect distributions. The post-rolled cooling treatments, such as water cooling, air cooling, and the reducing atmosphere as well as the cooling rate, all can affect the composition and microstructure of the final oxide scale [5, 8–12]. Chen and Yuen [13] have reviewed the high-temperature oxidation of iron and carbon steel in details, including the temperature control to reduce the oxide scale blistering, which is unnecessary to repeated here.

Table 4.1 Typical mill scales and their defects, reprinted from [8] by permission of Wiley–VCH

Scales	Reason	Explanation
Flaky	The finishing rolling temperature is high	Blistering and rolling-in of the secondary scale generated between stands
Sand-like	The roll surface is degraded	Satin-like roughening of the roll surface at the finishing stand and the rolled-in secondary scale
Meteor-like	The roll surface is degraded	Meteor-like roughening at the finishing stand and rolled-in secondary scale
Spindle-shaped	Imperfect descaling	Imperfect descaling in roughing and localized rolling in of the remaining scale
Red scale	Silicon scale	Imperfect descaling due to melting to above the eutectic point in the reheating furnace and wedge-like inclusions of fayalite into the underlying metal

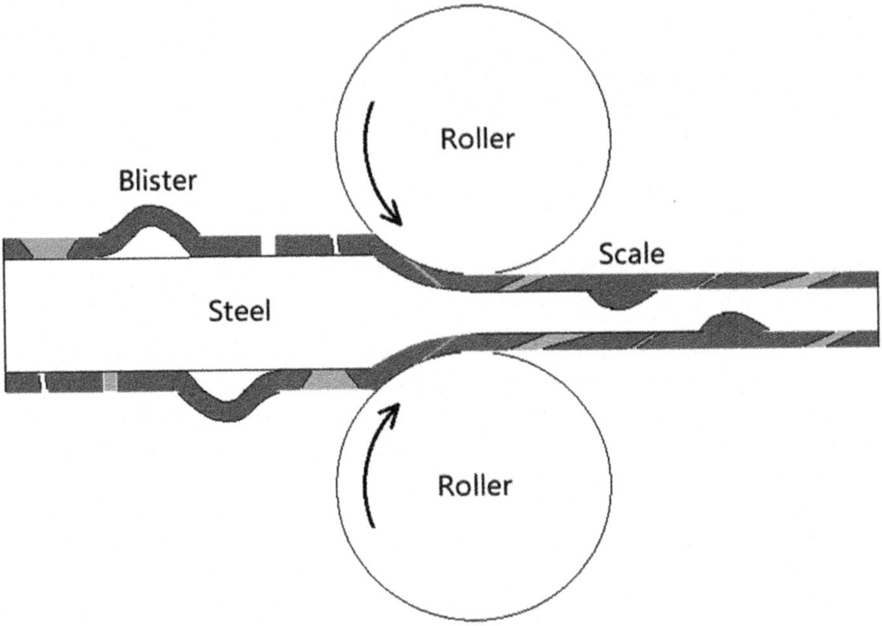

Fig. 4.7 Rolling-in defects of mill scale and steel surface

4.3 Microstructure of Rebar Mill Scale

4.3.1 Surface Morphology

Figure 4.8 is the electron microscope images of mill scale at the fracture zone and the rust stains of Φ8 mm HPB235 rebar [14]. Where there are no rust stains, the mill scale is blue-black, with a rough surface and micropores. At the rust stains, it is reddish-brown, with a loose and porous surface. There are "gullies" and "bulges" in Fig. 4.8a on the matrix surface along the rolling direction, and the matrix surface is rough and uneven.

Figure 4.9 is the surface photos of Φ12 mm HRB335 rebar mill scale [15]. The rolling imprints along the longitudinal direction of the rebar are clear in Fig. 4.9a, and the surface is rough and uneven, too. Figure 4.9b gives the enlarged mill scale surface morphology, which indicates large pores, longitudinal "gullies", and scaly cracks in the mill scale.

Figure 4.10 is the surface photos of Φ12 mm HRB400 rebar mill scale [16]. Similarly, the longitudinal rolling imprints can be clearly seen in Fig. 4.10a, and the surface is uneven. Figure 4.10b is the enlarged image of mill scale surface. Large holes and cracks are obvious, and some local mill scales are mushroom-shaped.

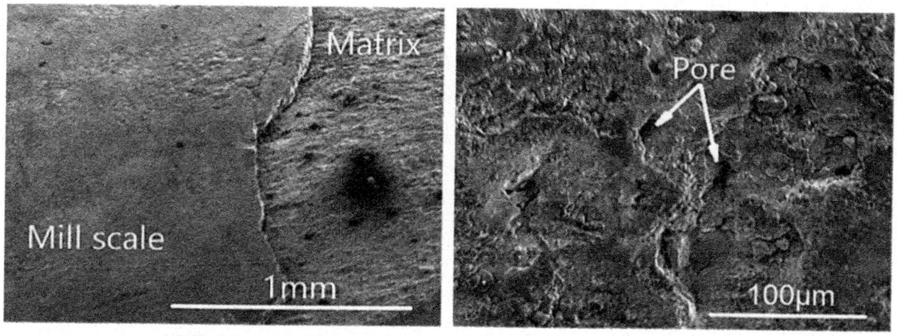

(a) Fracture zone (b) Stain spots

Fig. 4.8 Surface morphology of HPB235 rebar [14]

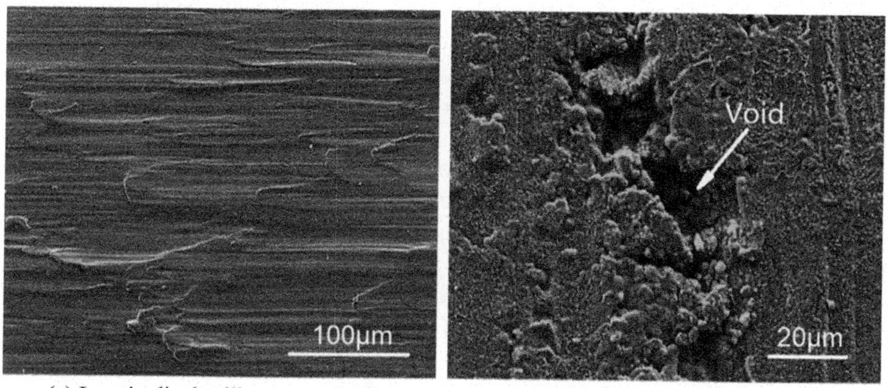

(a) Longitudinal rolling trace of rebar (b) Surface porosity of mill scale

Fig. 4.9 Surface morphology of HRB335 rebar [12]

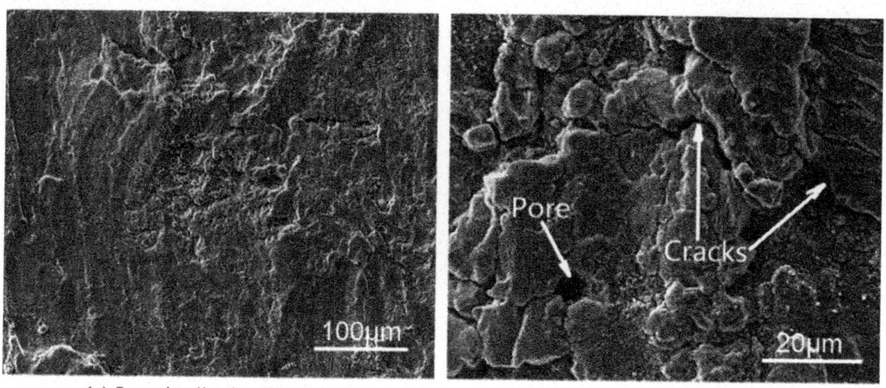

(a) Longitudinal rolling mark (b) Surface defects of mill scale

Fig. 4.10 Surface morphology of HRB400 rebar [16]

From Figs. 4.8, 4.9 and 4.10, it can be seen that although the surface morphologies of mill scales are different, there are micro defects such as pores and cracks in the mill scales, and the surfaces are rough and with obvious rolling imprints.

4.3.2 Section Morphology

4.3.2.1 Trilayer Structure

The mill scale of rebars oxidized at high temperature and with rapid cooling is usually a trilayer oxide scale. Figure 4.11 is the Φ8 mm HPB235 rebar section photos [14]. The HPB235 rebar mill scale is composed of FeO, Fe_3O_4, and Fe_2O_3 layers from inside to outside. There are cracks and tortuous interfaces between the rebar matrix and the FeO layer as well as the FeO and the Fe_3O_4 layer. The inner FeO layer is the loosest. The Fe_3O_4 layer is relatively dense. Figure 4.11b is a partially enlarged photograph of Fig. 4.11a. There are micro defects such as cracks and pores in the FeO layer.

Figure 4.12 is the oblique section photo of Φ8 mm HPB235 rebar [14]. Loose FeO layer and uneven rebar matrix surface can be seen more clearly (see the enlarged Fig. 4.12b). Fe_3O_4 layer in the mill scale is relatively dense, but there are some pores and cracks.

Since the thickness of Fe_2O_3 in the outermost layer is less than 1% of the total thickness [2], and the stratification with Fe_3O_4 is not particularly obvious, then the mill scale seems to be mainly divided into two parts, namely, Fe_3O_4 of dense outer layer and FeO of loose inner layer.

(a) HPB235 section (600×) (b) HPB235 section (1000×)

Fig. 4.11 Section image of HPB235 rebar [14]

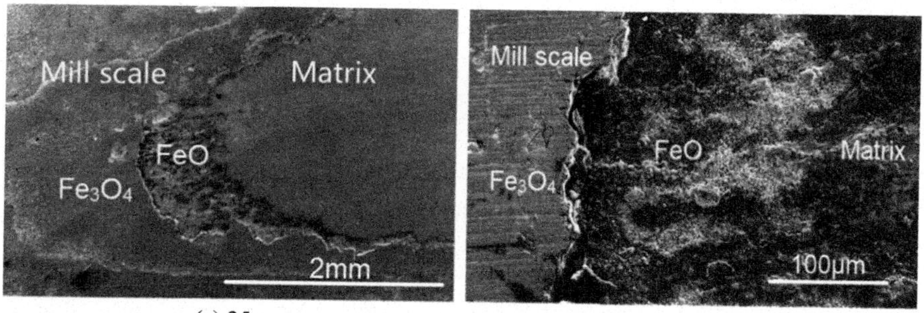

(a) 25×

(b) 300×

Fig. 4.12 Oblique section photo of HPB235 rebar [14]

4.3.2.2 Bilayer Structure

Figure 4.13 is the section photos of Φ12 mm HRB335 reinforcement [15]. The rebar mill scale is composed of FeO and Fe_3O_4 from inside to outside, mainly Fe_3O_4. The scale thickness is about 5–10 μm. There are lots of micro-defects, which are manifested in fractures and separation from the matrix, with a crack width varying from 0.5 to 8 μm.

Figure 4.14 is the section photos of Φ12 mm HRB400 reinforcement [16]. The mill scale is mainly composed of Fe_3O_4, and there is a small amount of FeO near the rebar matrix. The thickness of mill scale varies from 2 to 25 μm. There are micro defects such as cracks and pores in mill scale, and many cracks between the mill scale and the matrix, with a width of about 0.3–4 μm.

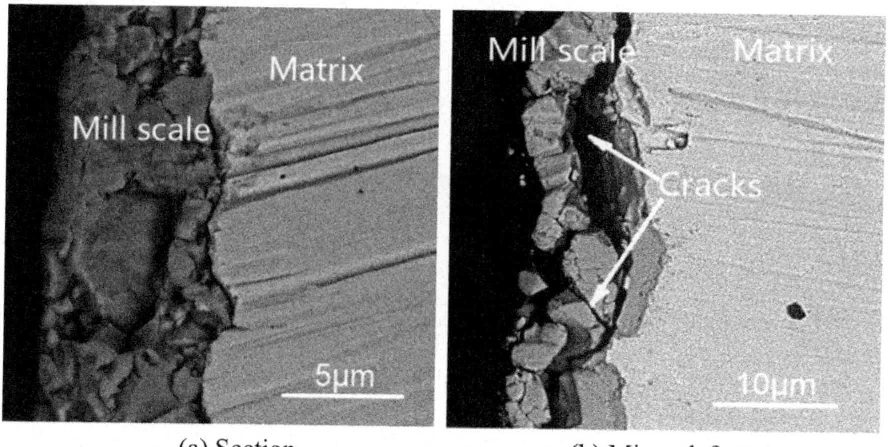

(a) Section

(b) Micro defects

Fig. 4.13 Section image of HRB335 rebar [15]

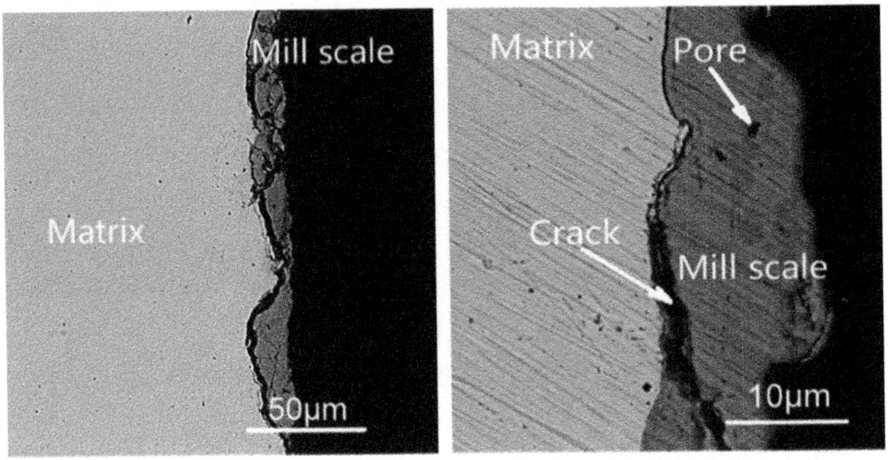

(a) Section (b) Micro defects

Fig. 4.14 Section image of HRB400 rebar [16]

It can be seen from the images above that different rebars have different mill scale compositions and microstructures. The dense mill scale may have a certain mechanical isolation effect which can delay the external corrosive medium penetrating into the matrix for a short period. However, the micro defects, especially the cracks, cannot prevent the invasion of the external media. When water and aggressive ions pass through the mill scale and contact with the rebar matrix, local corrosion of the substrate can be induced. The existence of microcracks always provides convenience for crevice corrosion of the rebar matrix.

4.4 Semiconductive Properties of Mill Scale

4.4.1 Conductivity of Iron Oxide

4.4.1.1 FeO

FeO (Ferrous oxide) is a covalent compound of Fe^{2+} and oxygen. It has a black appearance and with the mineral form of Wüstite (see Appendix B).

FeO has a cubic crystal structure, with 6 oxygen atoms around each iron atom and 6 iron atoms around each oxygen atom, forming the octahedral coordination, see Fig. 4.15 [17]. According to quantum physics and chemistry, one can know that the atomic arrangements and the electronic configurations determine the physical and chemical properties of the materials [18–20]. Therefore, different iron oxides have different conductivities under different conditions.

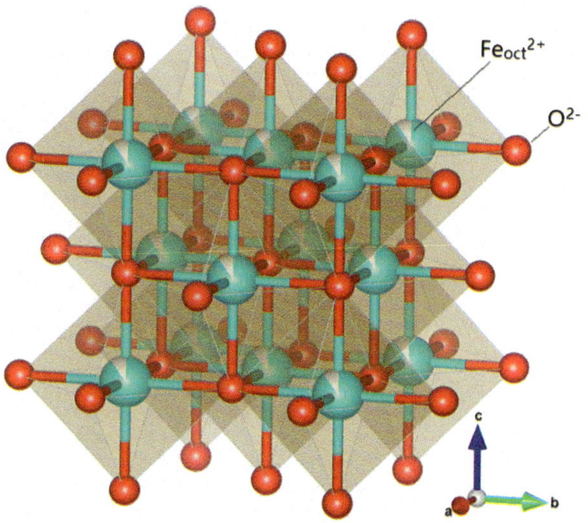

Fig. 4.15 Crystal structure of $Fe_{1-x}O$, according to the data from [17]

It can be seen from the Fe—O phase diagram in Fig. 4.1 [1] that above 570 °C, the phase equilibrium lines of Fe-FeO and FeO-Fe_3O_4 are curves, which means, FeO is not a stoichiometric compound, it is a typical metal deficient oxide, which is often recorded as $Fe_{1-x}O$, and its chemical composition can be from $Fe_{0.845}O$ to $Fe_{0.945}O$, which has high cation vacancy concentration and high mobility of cations and electrons, usually demonstrates the p-type semiconductor characteristics [2, 21].

The main reason why FeO is a non-stoichiometric compound is that Fe^{2+} is easily oxidized to Fe^{3+}, and Fe^{3+}, Fe^{2+} vacancies, and the electron holes are generated at the same time. Metallic vacancies can occupy tetrahedral or octahedral positions. When the vacancy concentration is high, vacancy clusters can form. The conductivity of FeO at high temperature (>570 °C) is mainly due to the jump of d electrons between the slightly overlapped electron orbits of Fe^{2+} and the nearest neighbor Fe^{3+}, while it is mainly conducted by the electron holes at low temperature [21].

4.4.1.2 Fe_3O_4

Fe_3O_4 (Ferroferric oxide), commonly known as iron oxide black, magnet, and black iron oxide, is a magnetic black crystal, also known as magnetic iron oxide, and the mineral is magnetite.

Fe_3O_4 can have a spinel or inverse spinel structure and may act as an n-type or p-type semiconductor under different conditions. In the Fe_3O_4 inverse spinel structure, i.e. $[Fe^{3+}]_{tet}[Fe^{2+}Fe^{3+}]_{oct}O_4$, O is cubic and closely packed, see Fig. 4.16. Fe^{2+} and Fe^{3+} at octahedral positions are paired, and the electrons can transfer rapidly between them at high temperatures. If conditions are met at low temperatures, the electrons can transmit through the tunneling effect, so Fe_3O_4 is a good electronic conductor [2, 21].

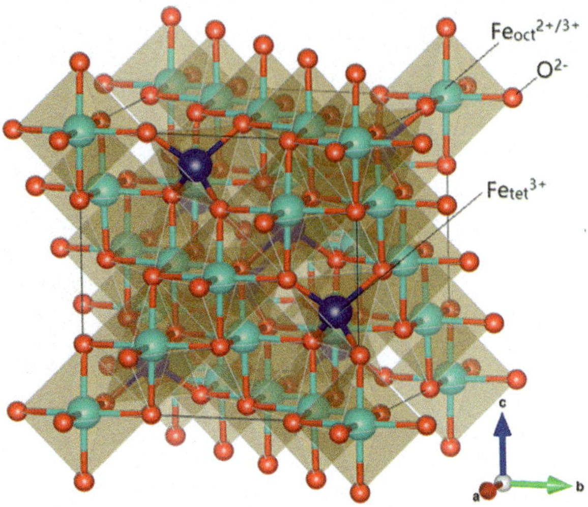

Fig. 4.16 Crystal structure of Fe_3O_4, according to the data from [17]

Fe_3O_4 will undergo metal—insulator transition at 120 K. It is called Verwey transition [22]. This transition reflects the ordering of Fe^{2+} and Fe^{3+} and transportable electrons [18, 23]. Fe_3O_4 can be written as $Fe^{3+}[Fe_2^{2.5+}]O_4$, that is, Fe^{3+} at two octahedral positions can be paired and share one electron. They can be affected by external electric or temperature fields. In essence, the Verwey transition is driven by Coulomb repulsion force between the adjacent electrons [18].

4.4.1.3 Fe_2O_3

Fe_2O_3 (Ferric oxide) is red or reddish-brown in appearance, known as iron red or iron oxide red, having an α-Fe_2O_3 or γ-Fe_2O_3 structure, and the corresponding minerals are hematite and maghemite, respectively.

Among the iron oxides, α-Fe_2O_3 is the closest stoichiometric compound, with a rhombic hexagonal crystal structure (corundum structure, same as Al_2O_3), see Fig. 4.17. Under some conditions, it can generate small oxygen-deficient n-type semiconductors, which can be written as Fe_2O_{3-y}, conducting through oxygen vacancy and Fe^{3+} interstitial ion migration [18, 21].

γ-Fe_2O_3 is a metastable compound with a cubic or tetragonal crystal structure, which is usually the same as that of Fe_3O_4, but there is only Fe^{3+} at the octahedral position, see Fig. 4.18. In order to maintain the electrical neutrality, it can be written as $[Fe_8^{3+}]_{tet}[Fe_{40/3}^{3+}V_{Fe8/3}]_{oct}O_{32}$, that is, the Fe vacancy appears only at the octahedral positions [18].

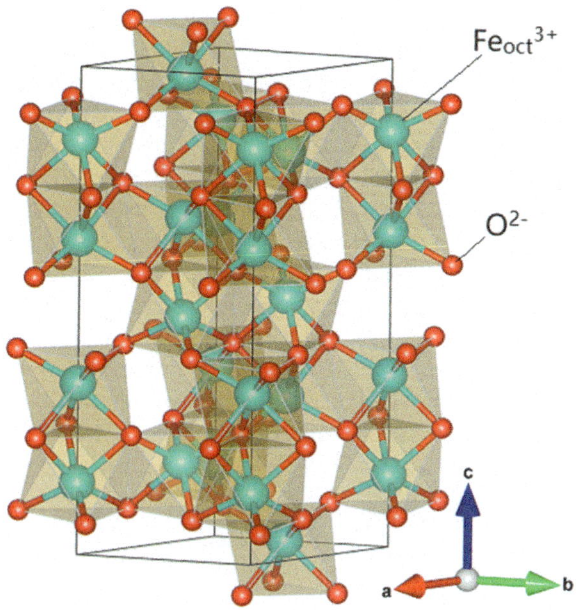

Fig. 4.17 Crystal structure of α-Fe_2O_3, according to the data from [17]

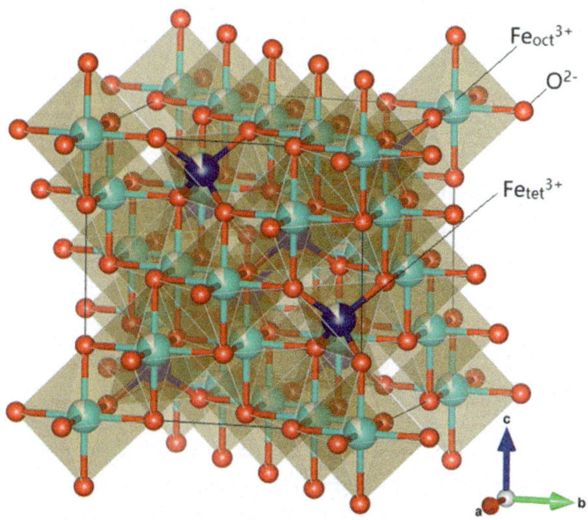

Fig. 4.18 Crystal structure of γ-Fe_2O_3, according to the data from [17]

4.4.2 Semiconductive Properties of Scale-Solution System

4.4.2.1 Mott-Schottky Curve

Because the Fermi energy level of semiconductor is different from that of solution, when the two phases contact, the Fermi energy levels of the semiconductor and the solution tend to be consistent, resulting in charge redistribution at the semiconductor/solution interface. When equilibrium reaches, a space charge layer will form at the interface, and the Helmholtz layer will be formed on the side of solution, then the electrode is formed.

The capacitance of the electrode surface is equal to the sum of the capacitance of the space charge layer and the Helmholtz layer in series. The capacitance of the space charge layer is much smaller than that of the Helmholtz layer, so the electrode surface capacitance is almost equal to that of the space charge layer. If the space charge layer is regarded as a flat plate capacitor, there is a Mott-Schottky relationship between its capacitance C_{cs} and the applied voltage E as shown in Eqs. (4.1) and (4.2). The curve drawn with E as the abscissa and $C_{cs}{}^{-2}$ as the ordinate is called the Mott-Schottky curve (abbreviated as MS), which is often used to analyze and judge the semiconductor type of passive film on the metal surface and calculate the donor concentration and the Flat band potential [24–29].

n-type semiconductor:

$$\frac{1}{C_{cs}^2} = \frac{2}{e\varepsilon_0\varepsilon_r N_D A^2}\left(E - E_{FB} - \frac{KT}{e}\right) \tag{4.1}$$

p-type semiconductor:

$$\frac{1}{C_{cs}^2} = -\frac{2}{e\varepsilon_0\varepsilon_r N_A A^2}\left(E - E_{FB} - \frac{KT}{e}\right) \tag{4.2}$$

where, ε_0 is the vacuum dielectric constant (8.854×10^{-14} F/cm); ε_r is the relative dielectric constant of semiconductor at room temperature; N_D and N_A are donor concentrations, respectively; A is the measured area; E is the applied voltage; E_{FB} is the Flat band potential; K is the Boltzmann constant (1.38×10^{-23} J/K); T is the thermodynamic temperature (K); e is the electron charge (1.602×10^{-19} C); KT/e is about 25 mV at room temperature.

4.4.2.2 Mill Scale-$Ca(OH)_2$ Solution

Figure 4.19 is the Mott-Schottky curve of mill scale peeled off using epoxy resin from the HRB400 rebar (so the tested surface is the inner surface of mill scale, mainly Fe_3O_4, see Fig. C-2d in Appendix C) after immersion in saturated $Ca(OH)_2$ solution for 15 min [16].

There are approximately three straight lines with a positive slope in the potential ranges of −0.6 – −0.2 V, −0.2 – 0.2 V, and 0.2 – 0.5 V, that is, the mill scale has an n-type semiconductive character in these potential ranges. When the potential is higher than 0.5 V, the slope of the curve becomes negative, indicating that the mill scale has a p-type semiconductive character. In this potential zone, water is electrolyzed and oxygen bubbles are releasing out the solution.

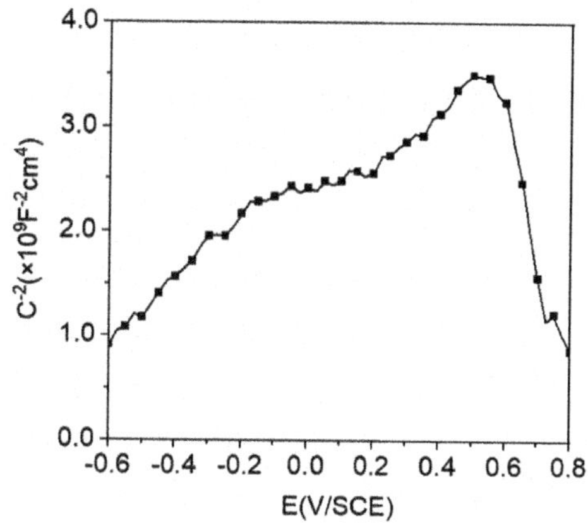

Fig. 4.19 Mott-Schottky curve of the HRB400 rebar mill scale in saturated $Ca(OH)_2$ solution [16]

4.5 Summary

The compositions, morphologies, and the microstructures of the black rebar with mill scales are briefly introduced. Plenty of microdefects are in the mill scale and at the scale/matrix interface. Pores and cracks provide convenient channels for water or corrosive ions to penetrate in. The mill scale-simulated concrete pore solution system usually demonstrates a n-type semiconductor characteristic in some potential ranges.

References

1. Wriedt, H.: The fe-O (Iron-Oxygen) system. JPE **12**, 170–200 (1991)
2. Birks, N., Meier, G.H., Pettit, F.S.: Introduction to the High-Temperature Oxidation of metals, 2nd edn. Cambridge University Press, Cambridge (2006)
3. Hauffe, K.: Scaling Processes in Metals and Alloys with Formation of Thick Protective Layers. In: Oxidation of Metals, pp. 273–288. Springer, Boston, MA (1965)
4. Cao, G., Wu, T., Xu, R., et al.: Effects of coiling temperature and cooling condition on transformation behavior of tertiary oxide scale. J. Iron Steel Res. Int. **22**, 892–896 (2015)
5. Picque, B.: Experimental study and numerical simulation of iron oxide scales mechanical behavior in hot rolling. École Nationale Supérieure des Mines de Paris. Ph.D. thesis (2004)
6. Yu, X.L., Jiang, Z.Y., Zhao, J.W., et al.: A review of microstructure and microtexture of tertiary oxide scale in a hot strip mill. Key Eng. Mater. **716**, 843–855 (2016)
7. Graf, M., Kawalla, R.: Scale development on steel during hot strip rolling. La Metallurgia Italiana. **2**, 43–49 (2014)
8. Krzyzanowski, M., Beynon, J.H., Farrugia, D.C.J.: Oxide Scale Behaviour in High Temperature Metal Processing. WILEY-VCH Verlag GmbH & Co, Weinheim (2010)
9. Sun, W.H.: A study on the characteristics of oxide scale in hot rolling of steel. University of Wollongong, Ph.D. thesis (2005)
10. Utsunomiya, H., Hara, K., Matsumoto, R., et al.: Formation mechanism of surface scale defects in hot rolling process. CIRP Ann. **63**(1), 261–264 (2014)
11. Ahmadi, D.: Oxide scales behaviour during descaling and hot rolling. University of Sheffield, Ph.D. thesis (2019)

12. Wei, J., Dong, J.H., Ke, W.: The influence of cooling processes on the corrosion performance of the rebar scale. Constr. Build. Mater. **24**, 275–282 (2010)
13. Chen, R.Y., Yuen, W.Y.D.: Review of the high-temperature oxidation of iron and carbon steels in air or oxygen. Oxid. Met. **59**(5/6), 433–468 (2003)
14. Lin, W.: Influences on rebar corrosion by rebar surface states, pH of concrete pore solution and [Cl-] in concrete. Tsinghua University. Master thesis (2002). In Chinese
15. Li, Y.J.: Chemical stability of rebar with mill scale in simulated concrete pore solutions. Tsinghua University. Master thesis (2016). in Chinese
16. Wang, X.: Redox of rebar with mill scale in simulated concrete pore solutions. Tsinghua University. Master thesis (2019). In Chinese
17. Villars, P. (Chief Editor): PAULING FILE in: Inorganic Solid Phases, SpringerMaterials (online database), Springer, Heidelberg (ed.). $Fe_{1-x}O$ (FeO ht) Crystal Structure, https://materials.springer.com/isp/crystallographic/docs/sd_0309017; Fe_3O_4 Crystal Structure, https://materials.springer.com/isp/crystallographic/docs/sd_0306168 ; α-Fe_2O_3 (Fe_2O_3 hem) Crystal Structure, https://materials.springer.com/isp/crystallographic/docs/sd_0381107 ; γ-Fe_2O_3 ($Fe_{2.67}O_4$ ht) Crystal Structure, https://materials.springer.com/isp/crystallographic/docs/sd_0558665 (2016)
18. Parkinson, G.S.: Iron oxide surfaces. Surf. Sci. Rep. **71**(1), 272–365 (2016)
19. Meng, Y., Liu, X.W., Huo, C.F., et al.: When density functional approximations meet iron oxides. J. Chem. Theory Comput. **12**(10),5132−5144 (2016)
20. Liao, P.: Mechanical, Optical, Transport, and Catalytic Properties of Iron Oxides from First Principles. Princeton University. Ph.D. dissertation (2012)
21. Gleitzer, C.: Electrical properties of anhydrous iron oxides. Key Eng. Mater. **125–126**, 355–418 (1996)
22. Verwey, E.: Electronic conduction of magnetite (Fe_3O_4) and its transition point at low temperatures. Nature **144**, 327–328 (1939)
23. Imada, M., Fujimori, A., Tokura, Y.: Metal-insulator transitions. Rev. Mod. Phys. **70**(4), 1039–1263 (1998)
24. Azumi, k., Ohtsuka, T., Sato, N.: Mott-schottky plot of the passive film formed on iron in neutral borate and phosphate solutions. J. Electrochem. Soc. **134**(6), 1352−1357 (1987)
25. Cheng, Y.F., Luo, J.L.: Electronic structure and pitting susceptibility of passive film on carbon steel. Electrochim. Acta **44**(17), 2947–2957 (1999)
26. Fujimoto, S., Tsuchiya, H.: Semiconductor properties and protective role of passive films of iron base alloys. Corros. Sci. **49**(1), 195–202 (2007)
27. Albery, W.J., O'shea, G.J., Smith, A.L.: Interpretation and use of mott-schottky plots at the semiconductor/electrolyte interface. J. Chem. Soc. Faraday Trans. **92**, 4083–4085 (1996)
28. La Mantia, F., Habazaki, H., Santamaria, M., et al.: A critical assessment of the Mott-Schottky analysis for the characterisation of passive film-electrolyte junctions. Russ. J. Electrochem. **46**, 1306–1322 (2010)
29. Gelderman, K., Lee, L., Donne, S.W.: Flat-band potential of a semiconductor: using the mott-schottky equation. J. Chem. Educ. **84**(4), 685–688 (2007)

Chapter 5
Passivation of Hot-Rolled Rebars

Abstract The pseudo-passivation of black rebar with mill scale in simulated concrete pore solutions is presented. It is revealed that the pseudo-passivation of hot-rolled rebar is the inward gradual oxidation of mill scale from the surface to inside.

5.1 Passivation of Hot-Rolled Rebars

5.1.1 HPB235

This chapter will discuss the passivation behavior of black rebars with mill scales.

Figure 5.1 is the anodic polarization curves of Φ8mm HPB235 rebar with mill scale in simulated concrete pore solutions with different pH values (test conditions are the same as Fig. 3.1) [1].

Different from the results of polished rebar in Fig. 3.1, the HPB235 reinforcement with mill scale has no typical passivation characteristic in simulated concrete pore solution, i.e., without typical "passive" potential region nor a typical "passive" current, but showing an obvious "pseudo-passivation" characteristic. In addition, with the increase of pH values, the curve moves to the lower right. The currents at 0.2 V and pH = 10.0 to 13.5 are 0.31 μA/cm^2, 0.36 μA/cm^2, 1.2 μA/cm^2 and 2.7 μA/cm^2, respectively, much less than the passive currents of the polished rebar (15–20 μA/cm^2 in Fig. 3.1).

Although it is uncertain whether any passive films have been already formed on the reinforcement, at least the physical isolation effect of mill scale cannot be ignored, even though the chemical stability of the mill scale decreases with the increase of pH values.

X. Lu, *Passivation and Corrosion of Black Rebar with Mill Scale*, Engineering Materials,
https://doi.org/10.1007/978-981-19-8102-9_5

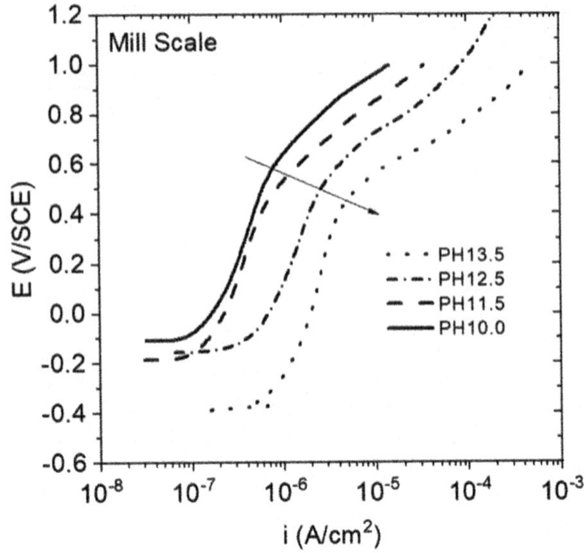

Fig. 5.1 Anodic polarization curves of HPB235 rebar under different pH

5.1.2 HRB335

Figure 5.2 is the anodic polarization curves of Φ12 mm HRB335 rebar with mill scale after natural immersion in saturated $Ca(OH)_2$ solution for 5 min and 140 days [2].

After immersion for 140 days, the anodic polarization curve of HRB335 rebar moves slightly to the upper left, the self-corrosion potential increases significantly,

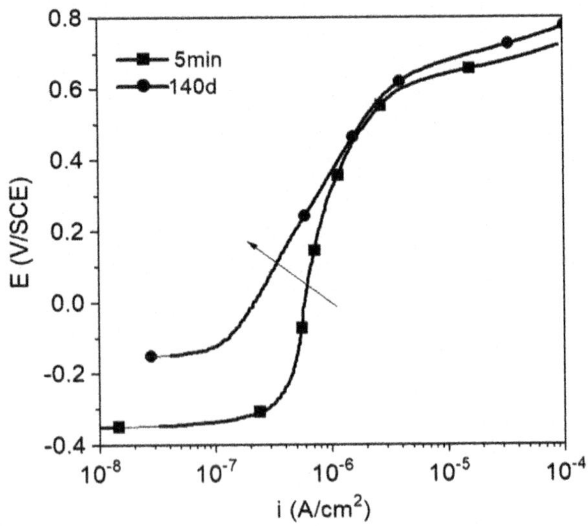

Fig. 5.2 Anodic polarization curves of HRB335 rebar after different immersion time

and the polarization curve demonstrates a "pseudo-passivation" characteristic. The anodic polarization current at 0.2 V is reduced from 0.78 μA/cm^2 at immersion of 5 min to 0.49 μA/cm^2, which is much smaller than the passive currents of the polished or the pickled one (6 μA/cm^2 and 12 μA/cm^2 in Fig. 3.2, respectively).

Although it is temporarily impossible to determine whether it has been already passivated, the physical isolation effect of the mill scale cannot be eliminated.

5.1.3 *HRB400*

Figure 5.3 is the multiple dynamic potential scanning curves of Φ12 mm HRB400 rebar with mill scale in saturated $Ca(OH)_2$ solution [3]. The first scan starts after immersion for 15 min, and the interval between the two scans is 2 h.

Although there is no typical passivation curve of HRB400 rebar with mill scale in saturated $Ca(OH)_2$ solution, the curve moves slightly to the upper left with the polarization scans. The anodic polarization current at 0.2 V decreases with the increase of scans: 0.54, 0.24, 0.15 and 0.14 μA/cm^2 in series. Though it is not possible to determine whether it has been passivated already as that shown in Fig. 3.3, it is certain that the surface of HRB400 rebar with mill scale has also been oxidized under the anodic polarization, resulting in the anodic polarization current remarkably decreases and the self-corrosion potential slightly increases.

Figure 5.4 is the potentiodynamic polarization curves of HRB400 rebar with mill scale after natural immersion in saturated $Ca(OH)_2$ solution for 15 min and 360 days [3]. Similarly, except that the polarization curve moves to the upper left, the rebar after long-term immersion still shows a "pseudo-passivation" characteristic as it starts, i.e.,

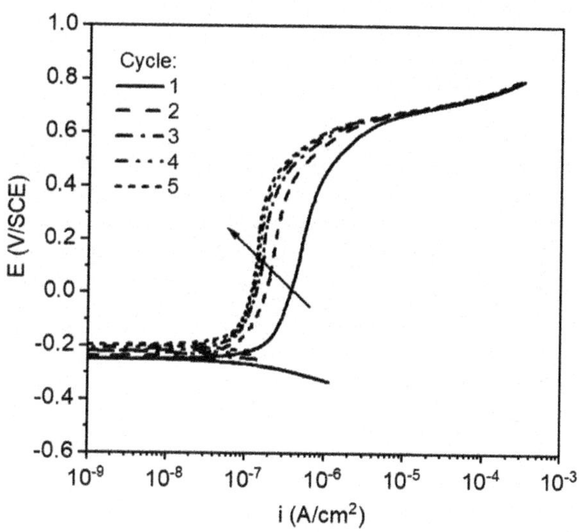

Fig. 5.3 Potentiodynamic scanning curves of HRB400 rebar

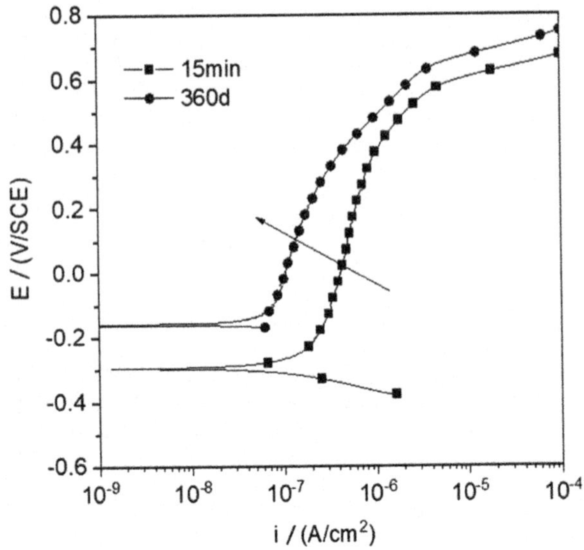

Fig. 5.4 Potentiodynamic scanning curves of HRB400 rebar after natural immersion different time

the electrode process has no obvious change, but the anodic polarization current at 0.2 V is reduced from 0.61 μA/cm^2 at immersion of 15 min to 0.19 μA/cm^2.

For the rebar with mill scale, with a small anodic polarization current (< 1 μA/cm^2) showing a "pseudo-passivation" behavior needs to be analyzed in combination with other methods to determine if there is any passive film formed or not.

5.2 Oxidation of Mill Scale

5.2.1 HRB335

Figures 5.5, 5.6 and 5.7 are the Raman spectra at different depths on the section of HRB335 rebar with mill scale as received, immersed in saturated $Ca(OH)_2$ solution for 140 days and potentiostatically polarized for 120 h at 0.35 V, respectively.

It can be seen that: (1) the mill scale of rebar as received is composed of Fe_3O_4 from inside to outside; (2) after natural immersion for 140 days, the interior of the mill scale is still Fe_3O_4, but γ-FeOOH appears at some sites on the rebar surface, and no new phase generates at the interface between the scale and the matrix; (3) after potentiostatic passivation for 120 h, the interior of the mill scale is still Fe_3O_4, but the surface and the top part of the mill scale becomes α-Fe_2O_3 dominant, that is, the mill scale of Fe_3O_4 is gradually oxidized from the outside to inside, but no new phase is formed at the interface of mill scale/rebar matrix.

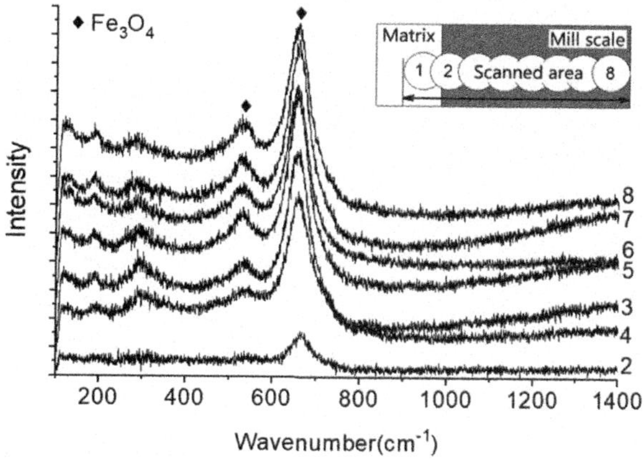

Fig. 5.5 Section Raman spectra of HRB335 rebar as received

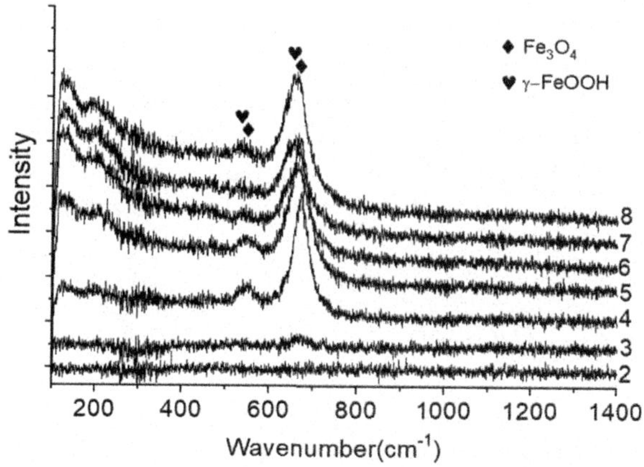

Fig. 5.6 Section Raman spectrum after natural immersion 140d

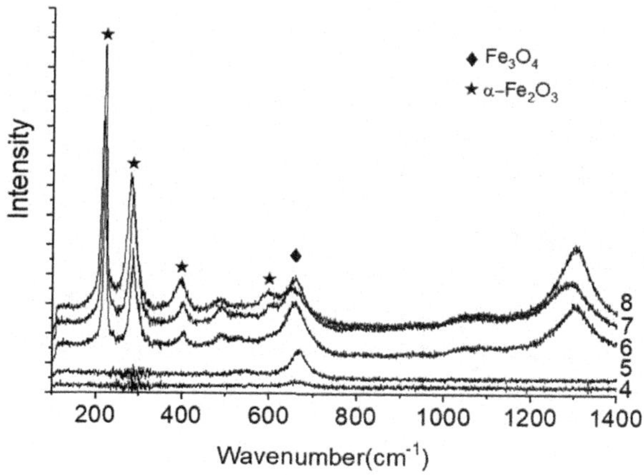

Fig. 5.7 Section Raman spectrum after potentiostatic passivation 120 h

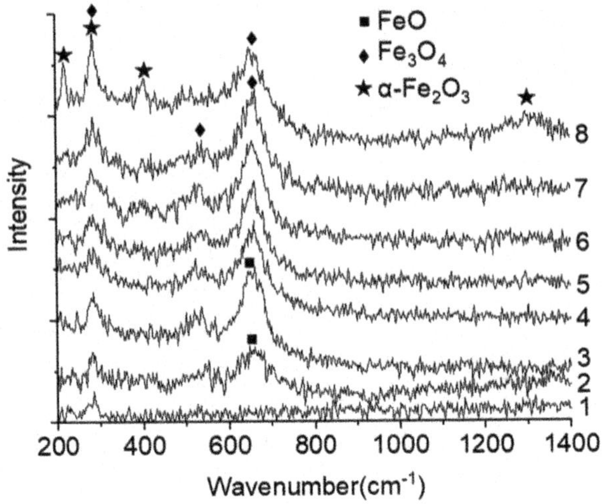

Fig. 5.8 Section Raman spectra of HRB400 rebar as received

Comparing Figs. 5.4 with 5.6, it can be deduced that the "pseudo-passivation" of HRB335 steel bar with mill scale is not the passivation of rebar matrix, but the inward oxidation of the mill scale from the surface to inside.

5.2.2 *HRB400*

Figures 5.8, 5.9 and 5.10 are the Raman spectra at different depths on the section of HRB400 rebar with mill scale as received, immersed in saturated $Ca(OH)_2$ solution for 360 days and the potentiostatically polarized for 60 days at 0.30 V, respectively.

It can be seen that: (1) the mill scale of the rebar as received is mainly composed of Fe_3O_4; there is small amount of FeO near the rebar matrix and α-Fe_2O_3 exists on the surface; (2) after natural immersion for 360 days, only α-Fe_2O_3 at the mill scale increases a bit, the other components do not change significantly; (3) after anodically polarized for 60 days, α-Fe_2O_3 at the mill scale surface increases while Fe_3O_4 decreases, which indicates the mill scale is gradually oxidized from the outside to inside (Fig. 5.9).

Figure 5.11 is the backscatter photos of the mill scale section of HRB400 rebars as received and anodically polarized 60 days [3]. The color depth at different positions of the mill scale of rebar as received is slightly different. The color at the outer side and the edge of pores is darker which means O content is higher.

After anodically polarized 60 days, the darker part of the mill scale extends from the surface to the inside, indicating that anodic polarization can gradually oxidize the mill scale from outside to inside. The darker area expands inward in the shape of spots and stripes, indicating the oxidation of mill scale is not uniform but along with the grain boundaries and other micro-defects.

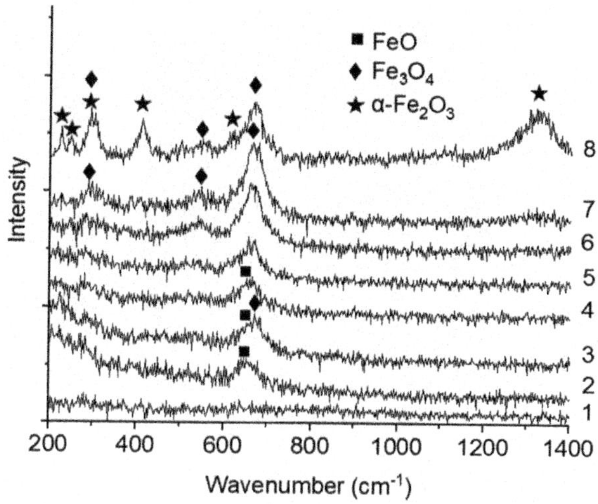

Fig. 5.9 Section Raman spectrum after natural immersion 360d

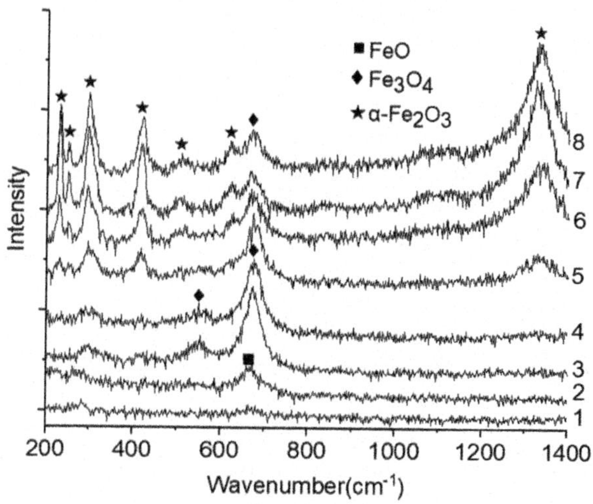

Fig. 5.10 Section Raman spectrum after anodic polarization 60d

5.3 Pseudo-Passivation Mechanisms of Rebar

According to above results, the pseudo-passivation of the hot-rolled black rebars in simulated concrete pore solutions can be regarded as the inward oxidation of mill scale from surface to inside, rather than the passivation of rebar matrix.

As shown in Fig. 5.12, taking the mill scale is composed of Fe_3O_4 as an example, when the rebar with mill scale is immersed in a strong alkaline solution without corrosive ions, OH^-, H_2O, and the dissolved oxygen [O] will first adsorb on the surface of the mill scale, and Fe^{2+} in Fe_3O_4 will lose another electron then turn to Fe^{3+}, which shows that local Fe_3O_4 on the surface of the mill scale is oxidized to

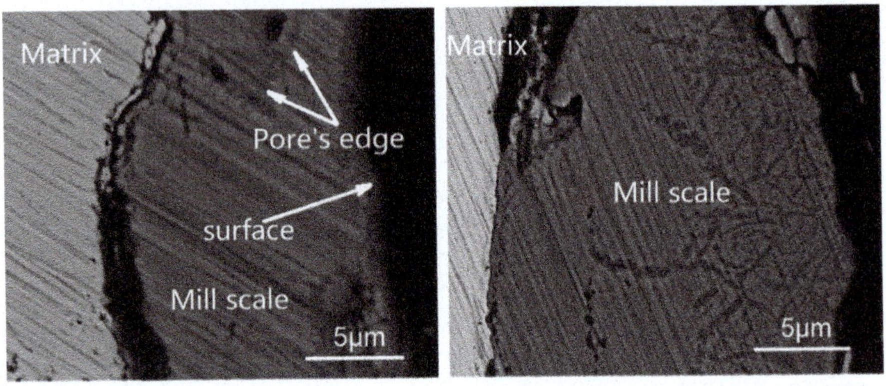

(a) Section as received (b) Section after anodic polarization 60d

Fig. 5.11 Section backscatter images of HRB400 as received and after anodic polarized for 60d

Fe_2O_3, and this reaction will gradually expand horizontally and vertically, then the inward oxidation of the mill scale from outside to inside takes place. It should be noted that Fe_3O_4 and Fe_2O_3 coexist at each section from the surface to the inside, although it is not strictly stratified, but the oxidized mill scale is dominated by Fe_2O_3.

Under strong anodic polarization, Fe_3O_4 can be oxidized to α-Fe_2O_3, and when the oxidizing condition is slightly weak, γ-Fe_2O_3 is formed, i.e., when the rebar is naturally immersed in saturated $Ca(OH)_2$ solution, Fe_3O_4 of the mill scale will usually transform to γ-Fe_2O_3 or γ-FeOOH.

Therefore, the tiny anodic polarization current of hot-rolled rebar in aqueous solution is considered mainly due to the mechanical isolation of the mill scale (significantly reducing the direct exposure area of the rebar matrix). It is not caused by the passive film formation on the rebar matrix, i.e., quite different from the descaled rebars as shown in Fig. 3.3.

Of course, the above reaction mechanism is just suitable for the dense mill scale. At those places with large defects, such as wide cracks in the mill scale, the strong alkaline solutions might contact the rebar matrix directly, whether any passive film

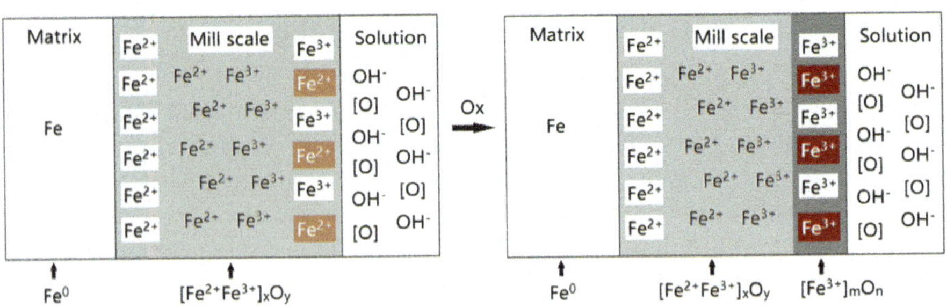

Fig. 5.12 Oxidation of mill scale from outside to inside

can form on the matrix surface needs to be discussed later. Despite the thermodynamic possibility, there is not any new phase has been observed yet at the matrix surface in our studies.

5.4 Summary

The pseudo-passivation of hot-rolled black rebar in strong alkaline solution is related to the inward gradual oxidation of the mill scale from the surface to inside. Fe_3O_4 can be oxidized to α-Fe_2O_3 under strong oxidizing conditions, such as strong anodic polarization, or to γ-Fe_2O_3 or γ-FeOOH under not strong enough condition, such as natural immersion in simulated concrete pore solutions.

References

1. Lin, W.: Influences on rebar corrosion by rebar surface states, pH of concrete pore solution and [Cl^-] in concrete. Tsinghua University. Master thesis (2002). In Chinese
2. Li, Y.J.: Chemical stability of rebar with mill scale in simulated concrete pore solutions. Tsinghua University. Master thesis (2016). In Chinese
3. Wang, X.: Redox of rebar with mill scale in simulated concrete pore solutions. Tsinghua University. Master thesis (2019). In Chinese

Chapter 6
Redox Reactions of Hot-Rolled Rebars

Abstract The reduction of the rebar mill scale is verified. The redox reactions of mill scale in simulated concrete pore solution are discussed according to the cyclic voltammetry curves and Raman spectra. The oxidation and reduction of hot-rolled rebar are mainly controlled by the redox of the mill scale.

6.1 Electrode Reactions of Fe in Water

Before discussing the redox reactions on hot-rolled rebars, it's better to understand the electrode reactions of Fe or descaled rebars in aqueous solutions, especially in alkaline conditions.

Table 6.1 lists the electrode reactions of Fe in H_2O and the thermodynamic equilibrium potential calculating equations [1], which are used to plot the E-pH diagram [2].

It should be noted that the electrode reactions of Fe in actual solutions will vary according to the solution composition, temperature, pH, etc. The actual reaction process may involve complex processes, such as adsorption, oxidation/reduction, reactant deposition, dehydration, etc. If there is lack of thermodynamic data on these intermediate reactions, it is difficult to carry out the theoretical calculations.

6.2 Reactions of Fe and Descaled Rebar in Alkaline Solutions

In practice, cyclic voltammetry curves (see the following section) are often employed to study the electrode reactions and their reversibility of metallic electrodes in different aqueous solutions.

Table 6.2 lists the electrode reactions of pure iron and the descaled black rebar in alkaline solutions together with their corresponding potential peaks in the cyclic voltammetry curves.

X. Lu, *Passivation and Corrosion of Black Rebar with Mill Scale*, Engineering Materials,
https://doi.org/10.1007/978-981-19-8102-9_6

Table 6.1 Electrode reactions and equilibrium potential calculating equations for Fe in H_2O, reprinted from [1] with permission of Springer Nature

Reactions	Equilibrium potential calculation formula
$Fe(s) \rightarrow Fe^{2+}(aq) + 2e$	$E = -0.440 - 0.0295 \log[Fe^{2+}]$
$3Fe(s) + 4H_2O(l) \rightarrow Fe_3O_4(s) + 8H^+(aq) + 8e^-$	$E = -0.085 - 0.0591pH$
$3Fe^{2+}(aq) + 4H_2O(l) \rightarrow Fe_3O_4(s) + 8H^+(aq) + 2e^-$	$E = 0.980–0.2364pH–0.0886 \log[Fe^{2+}]$
$2Fe^{2+}(aq) + 3H_2O(l) \rightarrow Fe_2O_3(s) + 6H^+(aq) + 2e^-$	$E = 0.728–0.1773pH–0.0591 \log[Fe^{2+}]$
$2Fe_3O_4(s) + 4H_2O(l) \rightarrow 2Fe_2O_3(s) + 2H^+(aq) + 2e^-$	$E = 0.221–0.0591pH$
$Fe(s) + 2H_2O(l) \rightarrow HFeO_2^-(aq) + 3H^+(aq) + 2e^-$	$E = 0.493–0.0886pH + 0.0295 \log[HFeO_2^-]$
$3HFeO_2^-(aq) + H^+(aq) \rightarrow Fe_3O_4(s) + 2H_2O(l) + 2e^-$	$E = -1.819 + 0.0295pH - 0.0886 \log[HFeO_2^-]$
$Fe^{2+}(aq) \rightarrow Fe^{3+}(aq) + e^-$	$E = 0.771 + 0.0591 \log([Fe^{3+}]/[Fe^{2+}])$
$2Fe^{3+}(aq) + 3H_2O(l) \rightarrow Fe_2O_3(s) + 6H^+$	$\log[Fe^{3+}] = -0.72 - 3pH$

The interpretation of the same reaction peak is not completely consistent. However, most people think that the peak I and II are corresponding to the oxidation of $Fe^0 \rightarrow Fe^{2+}$, which is mostly considered to be the process of $Fe \rightarrow Fe(OH)_2$; while peak III and III' are corresponding to the oxidation of $Fe^{2+} \rightarrow Fe^{3+}$, peak IV and IV' are the inverse reactions of III and III', respectively, and the peak V is the inverse reaction of peak I and II.

In the passive zone, Fe_3O_4 can be transformed to γ-FeOOH or γ-Fe_2O_3 [12]. When the potential is higher than 0.6 V, oxygen will be released and high valent iron ions could be generated [3–6, 8, 9]; while when the potential is lower than −1.1 V, hydrogen can be released, and part of the oxides may be reduced to zero-valent iron.

6.3 Cyclic Voltammetry of Rebar with Mill Scale

6.3.1 HRB335

Figures 6.1, 6.2 and 6.3 are cyclic voltammetry (CV) curves of HRB335 rebar with and without mill scale in saturated $Ca(OH)_2$ solutions [13] (①, ② in figurers indicates the first and the second cycle; the arrow indicates the increasing direction of the scanning cycles; and the same as below).

The cyclic voltammetry curves of HRB335 rebar with mill scale immersed in saturated $Ca(OH)_2$ solution for 5 min are remarkably different from those for 3 days. In Fig. 6.1, there is an oxidation peak at −0.25 V and a reduction peak at −1.4 V. In Fig. 6.2, there are two oxidation peaks at −0.90 and −0.55 V and the corresponding

Table 6.2 Reaction peaks and reactions of Fe and descaled rebar in strong alkaline solutions

Peak	Potential (V/SCE)	Metal and solution	Reaction
I	−1.1	Fe, Saturated $Ca(OH)_2$	$Fe \rightarrow Fe(OH)_2$ [3]
	−1.15	Fe, 1 M KOH	$Fe \rightarrow HFeO_2^-$ [4]
	−1.1	Fe, 1 M NaOH	$Fe + 2OH^- \rightarrow Fe(OH)_2 + 2e$ [5]
	−1.07	Fe, 1 M KOH	$Fe + OH^- \rightarrow [Fe(OH)]_{ads} + e$, $[Fe(OH)]_{ads} + OH^- \rightarrow [Fe(OH)_2]_{ads} + e$ [6]
	−1.05	Rebar, Saturated $Ca(OH)_2$	$Fe \rightarrow Fe(OH)_2$ [7]
	−0.98	Rebar, 0.8 M NaOH	$Fe + OH^- \rightarrow [Fe(OH)]_{ads} + e$, $[Fe(OH)]_{ads} \rightarrow [Fe(OH)]^+{}_{ads} + e$ [8]
	−1.1	HRB400, Saturated $Ca(OH)_2$	$Fe + 2OH^- \rightarrow Fe(OH)_2 + 2e$ [9]
II	−0.9	Fe, Saturated $Ca(OH)_2$	$Fe \rightarrow Fe(OH)_2$ [10]
	−0.95	Fe, 1 M KOH	$Fe \rightarrow Fe(OH)_2$[11]
	−1.0	Fe, 1 M NaOH	$Fe + 2OH^- \rightarrow Fe(OH)_2 + 2e$ [7]
	−0.91	Fe, 1 M KOH	$[Fe(OH)_2]_{ads} + 2OH^- \rightarrow 2Fe(OH)_2 + 2e$ [8]
	−0.93	Rebar, 0.8 M NaOH	$[Fe(OH)]^+{}_{ads} + 2OH^- \rightarrow HFeO_2^- + H_2O \rightarrow Fe(OH)_2 + OH$ [9]
III	−0.7	Fe, Saturated $Ca(OH)_2$	Hydrous outer oxide layer, Fe(II) → Fe(III) [3]
	−0.75	Fe, 1 M KOH	$Fe(OH)_2 \rightarrow \alpha, \delta, \delta'$-FeOOH[4]
	−0.91	Fe, 1 M NaOH	$3Fe(OH)_2 + 2OH^- \rightarrow Fe_3O_4 + 4H_2O + 2e$ [7]
	−0.55	Fe, 1 M KOH	$Fe(OH)_2 + OH^- \rightarrow FeOOH + H_2O + e$ [8]
	−0.52	Rebar, Saturated $Ca(OH)_2$	Porous outer layer, $Fe(OH)_2 \rightarrow Fe_3O_4$[5]
	−0.76	Rebar, 0.8 M NaOH	$Fe(OH)_2 + OH^- \rightarrow [FeOOH]_{ads} + H_2O + e$ [9]
	−0.6	HRB400, Saturated $Ca(OH)_2$	$3Fe(OH)_2 + 2OH^- \rightarrow Fe_3O_4 + 4H_2O + 2e$ [9]
III'	−0.5	Fe, Saturated $Ca(OH)_2$	Anhydrous inner oxide layer, Fe(II) → Fe(III) [3]
	−0.72	Fe, 1 M NaOH	$Fe_3O_4 + H_2O + OH^- \rightarrow 3FeOOH + e$[5]

(continued)

Table 6.2 (continued)

Peak	Potential (V/SCE)	Metal and solution	Reaction
	−0.3	Rebar, Saturated $Ca(OH)_2$	Dense inner layer, Fe(II) → Fe(III) [5]
	−0.71	Rebar, 0.8 M NaOH	$2[FeOOH]_{ads} \rightarrow Fe_2O_3 \cdot H_2O$ or $[FeOOH]_{ads} + H_2O \rightarrow FeOOH \cdot H_2O$ [9]
IV'		Fe, Saturated $Ca(OH)_2$	Anhydrous inner oxide layer, Fe(III) → Fe(II) [3]
	−0.95	Fe, 1 M NaOH	$3FeOOH + e \rightarrow Fe_3O_4 + H_2O + OH^-$[5]
	−0.63	Rebar, Saturated $Ca(OH)_2$	Dense inner layer, Fe(III) → Fe(II) [5]
	−1.0	Rebar, 0.8 M NaOH	$Fe_2O_3 \cdot H_2O \rightarrow 2[FeOOH]_{ads}$ or $FeOOH \cdot H_2O \rightarrow [FeOOH]_{ads} + H_2O$ [9]
IV	−1.0	Fe, Saturated $Ca(OH)_2$	Hydrous outer oxide layer, Fe(III) → Fe(II) [3]
	−1.1	Fe, 1 M KOH	Fe(III) → Fe(II) [3]
	−1.23	Fe, 1 M NaOH	$Fe_3O_4 + 4H_2O + 2e \rightarrow 3Fe(OH)_2 + 2OH^-$[5]
	−1.09	Fe, 1 M KOH	$FeOOH + H_2O + e \rightarrow Fe(OH)_2 + OH^-$ [8]
	−1.1	Rebar, Saturated $Ca(OH)_2$	Porous outer layer, $Fe_3O_4 \rightarrow Fe(OH)_2$[5]
	−1.09	Rebar, 0.8 M NaOH	$[FeOOH]_{ads} + H_2O + 2e \rightarrow Fe(OH)_2 + OH^-$ [9]
	−1.1	HRB400, Saturated $Ca(OH)_2$	$Fe_3O_4 + 4H_2O + 2e \rightarrow 3Fe(OH)_2 + 2OH^-$ [12]
V	−1.3	Fe, Saturated $Ca(OH)_2$	$Fe(OH)_2$ reduction [3]
	−1.25	Fe, 1 M KOH	Fe(II) → Fe^0 [4]
	−1.3	Fe, 1 M NaOH	$Fe(OH)_2 + 2e \rightarrow Fe + 2OH$[5]
	−1.26	Fe, 1 M KOH	$Fe(OH)_2 + 2e \rightarrow Fe + 2OH^-$ [8]
	−1.35	Rebar, Saturated $Ca(OH)_2$	Fe(II) → Fe^0[5]
	−1.18	Rebar, 0.8 M NaOH	$Fe(OH)_2 + 2e \rightarrow Fe + 2OH^-$ [9]
	−1.4	HRB400, Saturated $Ca(OH)_2$	$Fe(OH)_2 + 2e \rightarrow Fe + 2OH^-$ [12]
Passive region	−0.2 – 0.5	HRB400, Saturated $Ca(OH)_2$	$Fe_3O_4 + H_2O + OH^- \rightarrow 3\gamma\text{-}FeOOH + e$, $2Fe_3O_4 + 2OH^- \rightarrow 3\gamma\text{-}Fe_2O_3 + H_2O + 2e$ [12]

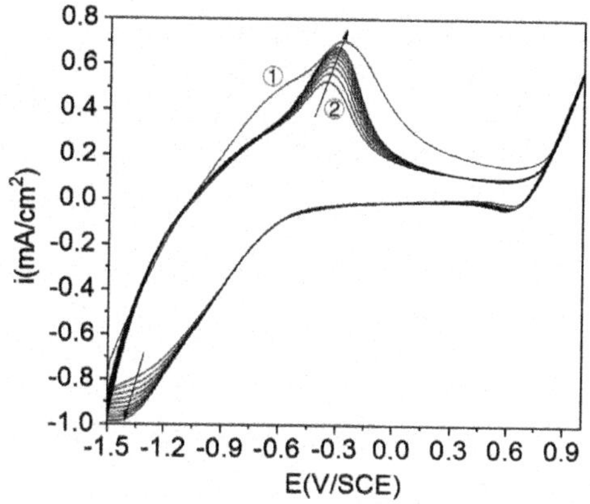

Fig. 6.1 Cyclic voltammetry curve of HRB335 rebar will mill scale immersed 5 min

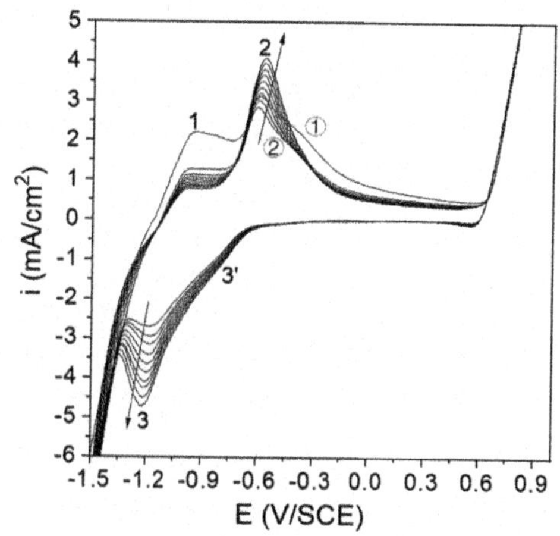

Fig. 6.2 Cyclic voltammetry curve of HRB335 rebar with mill scale immersed 3d

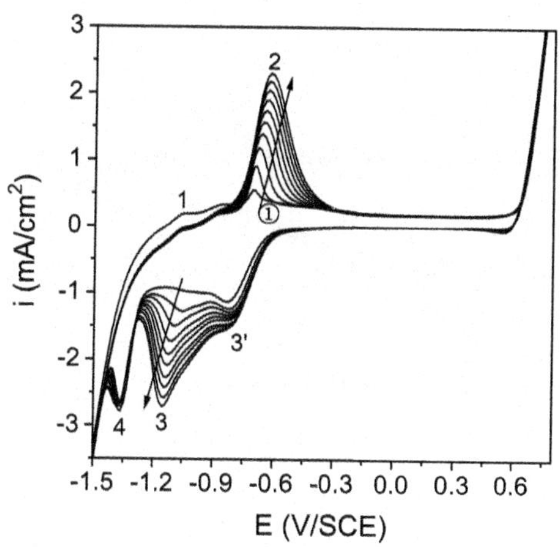

Fig. 6.3 Cyclic voltammetry curve of HRB335 rebar without mill scale immersed 5 min

currents increase with cycles. A reduction peak is at −1.2 V as well. Peak 2 and peak 3 are obviously asymmetric.

There are obvious passivation plateaus in the cyclic voltammetry curves of HRB335 rebars with and without mill scale (see Figs. 6.2 and 6.3). The "passive" potential region of rebar with mill scale is 0–0.6 V and that of rebar without mill scale is −0.25 – 0.60 V, indicating the latter one is easier to passivate.

It is generally believed that peak 1 corresponds to the oxidation of $Fe^0 \rightarrow Fe^{2+}$ (see Fig. 6.3), and peak 2 corresponds to the oxidation of $Fe^{2+} \rightarrow Fe^{3+}$. Compare Figs. 6.1 with 6.3, when HRB335 rebar with mill scale is immersed in solution after 5 min, its oxidation peak and reduction peak shift significantly positive and negative after the first scan, and there is no peak 1 in Fig. 6.1, i.e., there is no $Fe^0 \rightarrow Fe^{2+}$ reaction, and no hydrogen evolution reaction at −1.5 V; the small currents indicate that the mill scale is a poor conductor of electrons at the initial immersion stage and has a certain barrier effect. The existence of peak 2 indicates that Fe^{2+} in the mill scale can be oxidized. After immersion 3 days (see Fig. 6.2), the cyclic voltammetry curve of rebar with mill scale is somewhat similar to that of the rebar without mill scale, indicating that the solution has been in contact with the rebar matrix through the mill scale. Compare Fig. 6.3 with Fig. 6.2, it is not difficult to see that peak 1 of the descaled HRB335 rebar is not obvious, and there is peak 4 at −1.35 V which is indicating the reaction of $Fe^{2+} \rightarrow Fe^0$, but the same reduction peak is not observed in Fig. 6.2.

Figures 6.4 and 6.5 are the cyclic voltammetry curves of HRB335 rebars with scale, polished, and descaled when the upper limit of cycle is changed. It can be seen that when the upper limit is reduced to 0.0 V in Fig. 6.4a, the reduction peak 4 of the rebar with mill scale appears. When the upper limit changes to −0.8 V in Fig. 6.4b (i.e. peak 2 is excluded), peak 3 disappears, indicating peak 3 is related to peak 2 and corresponding to the reaction of $Fe^{3+} \rightarrow Fe^{2+}$.

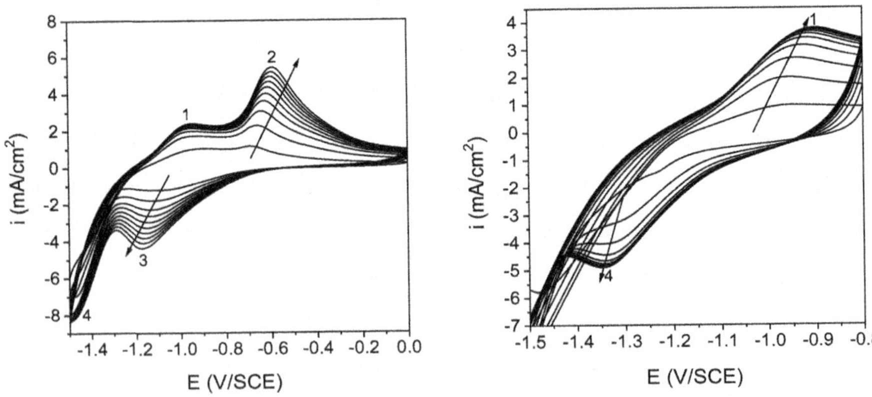

(a) Upper limit to 0.0V (b) Upper limit to -0.8V

Fig. 6.4 Cyclic voltammetric curves of HRB335 rebar with mill scale after the positive limit changed

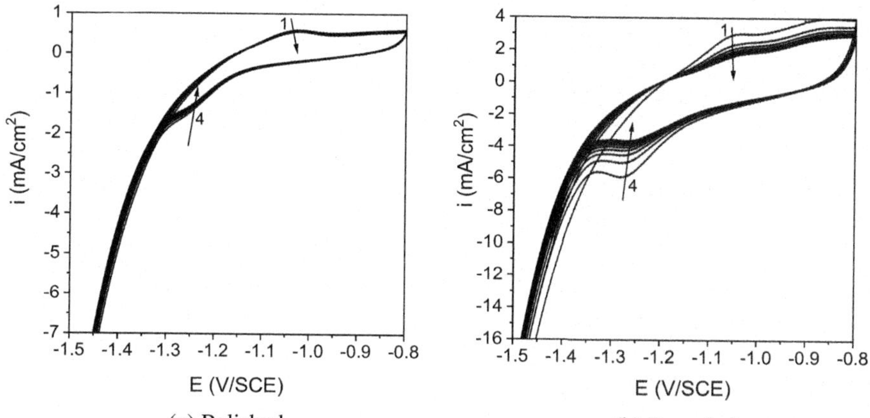

Fig. 6.5 Cyclic voltammetric curves of polished and descaled HRB335 rebar when the positive limit changes to −0.8 V

The peak currents of oxidation peak and reduction peak of rebar with mill scale in Fig. 6.4 increase with the scanning cycles, which is opposite in Fig. 6.5, indicating that Fe^{2+} accumulates at the interface of rebar matrix.

Compare with Fig. 6.2, it can be deduced that the reduction peak 4 only appears when the positive limit of cyclic potential does not get into the passivation plateau, which indicates that after entering the passivation plateau, the oxidation reaction does not take place at the mill scale/rebar matrix interface, but the $Fe^{2+} \rightarrow Fe^{3+}$ occurs in the mill scale instead. Otherwise, peak 4 should appear in Fig. 6.2. During the reduction, due to the sufficient supply of Fe^{2+} on the mill scale/rebar matrix interface, the polarization current does not decrease, and directly enters into the hydrogen evolution reaction.

Compare Fig. 6.4a and b, after the upper limit is further reduced, the current density of peak 4 gradually increases and the redox products accumulate, indicating that the repeated and rapid cathodic polarizations destroy the integrity of Fe^{2+} at the matrix surface and the inner layer of the mill scale, then Fe^{2+} at this interface is reduced, so that peak 4 appears.

It seems that the whole process is controlled by the redox reactions at the two interfaces, i.e. the $Fe^{2+} \leftrightarrow Fe^{3+}$ at the mill scale/solution interface and the $Fe^{0} \leftrightarrow Fe^{2+}$ at the mill scale/rebar matrix interface.

Figure 6.5 shows that the polished rebar, without any mill scale, has a very good reproducibility in redox reactions. Comparatively, the peak current of the descaled rebar is greater than that of the polished one and decreases with the increase of scanning cycles, which indicates the activity of the rough surface is much higher at the very beginning.

6.3.2 *HRB400*

Figure 6.6 is the cyclic voltammetry curves of Φ12 mm HRB400 descaled rebar after immersion in saturated $Ca(OH)_2$ solution for 15 min [14]. Here, the first scan is from the self-corrosion potential to −1.5 V, then from −1.5 to 0.5 V, and reverse scan to self-corrosion potential. In such a way, a cycle is completed, and the scanning rate is 5 mV/s.

As seen in Fig. 6.6, two smaller and relatively stable current peaks A1 and A2 first appear in the anodic scan. Then follows a peak A3 with a large current. The current of peak A3 increases with cycles, and the peak position moves forward positively. There is a shoulder peak A3' on the right side of peak A3. The peak A3' current is large in the first cycle, then decreases and remains stable. After A3', there is a wide passivation zone (−0.05 – 0.5 V), and the current is small, so an oxide film should be formed. A reduction peak C1 appears in the cathodic scan. The current of the peak C1 increases with the cycles, and its position shifts negatively. The electrode potential, peak position changes, and the electrode reactions corresponding to different redox peaks of the descaled HRB400 rebar are summarized in Table 6.3.

Figure 6.7 is the cyclic voltammetry curves of Φ12mm HRB400 rebar with mill scale after immersion in saturated $Ca(OH)_2$ solution for 15 min [14]. It can be seen that there are also peaks A1, A2, A3, A3', and C1, indicating that the electrode reaction is similar to that of descaled rebar. There is a reduction peak C2 at −1.3 V, but it is absent for the descaled rebar, so peak C2 should be related to the reduction of mill scale. All the peak currents increase with cycles, and all the oxidation peaks move positively together with the reduction peaks move negatively, which are obviously different from the behavior of the descaled rebar, indicating that the mill scale affects the electrode reactions.

Figure 6.8 is the cyclic voltammetry curves of the mill scale peeled from HRB400 rebar using epoxy and after immersion in saturated $Ca(OH)_2$ solution for 15 min. It should be noted that the testing surface is the inner surface of the mill scale. It can be

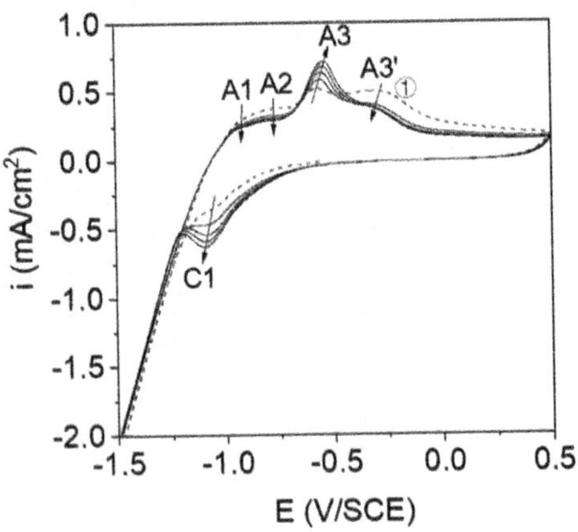

Fig. 6.6 Cyclic voltammetry curves of descaled HRB400 rebar

Table 6.3 Reactions corresponding to redox peaks and passive zone of descaled HRB400 rebar

Peak	Potential (V/SCE)	Change with cycles	Peak current changes with cycles	Electrode reaction
A1	−0.92	–	Slightly ↓	$Fe^0 \rightarrow Fe^{2+}$, $Fe \rightarrow Fe(OH)_2$ [3–6, 8, 9, 12]
A2	−0.79	–	Slightly ↓	
A3	−0.57	→	↑	$Fe^{2+} \rightarrow Fe^{3+}$, generate Fe_3O_4 [5, 6, 12]
A3'	−0.31	←	First ↓, then stable	Fe^{2+}, $Fe_3O_4 \rightarrow Fe^{3+}$ [3–5]
C1	−1.0	←	↑	A3, A3' reverse process, $Fe^{3+} \rightarrow Fe^{2+}$ [3–6, 8, 9, 12]
Passive region	−0.05 – 0.5			$Fe_3O_4 + H_2O + OH^- \rightarrow 3\gamma\text{-}FeOOH + e$ $2Fe_3O_4 + 2OH^- \rightarrow 3\gamma\text{-}Fe_2O_3 + H_2O + 2e$ [12]

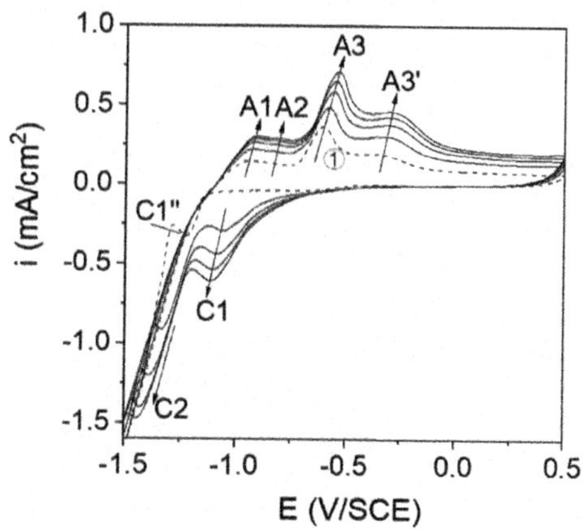

Fig. 6.7 Cyclic voltammetry curves of HRB400 rebar with mill scale

seen that the shape of cyclic voltammetry curve of the mill scale sample is similar to that of the rebar with mill scale, and the number of redox peaks as well as the current changing trends are similar, indicating that the electrode process of rebar with mill scale is mainly controlled by the redox reaction of the mill scale.

Comparing Figs. 6.6 with 6.7 and 6.8, the electrode reaction of HRB400 rebar is analyzed as follows:

① **Reaction of $Fe^0 \rightarrow Fe^{2+}$**: peak A1 and A2 appear during forward scan of rebar with mill scale and the mill scale sample, indicating that $Fe^0 \rightarrow Fe^{2+}$ reaction occurs. There may be three sources of Fe^0 in the mill scale sample: steel matrix remains, interstitial atoms in mill scale, or iron atoms reduced from mill scale during cathodic polarization at −1.3 – −1.5 V;

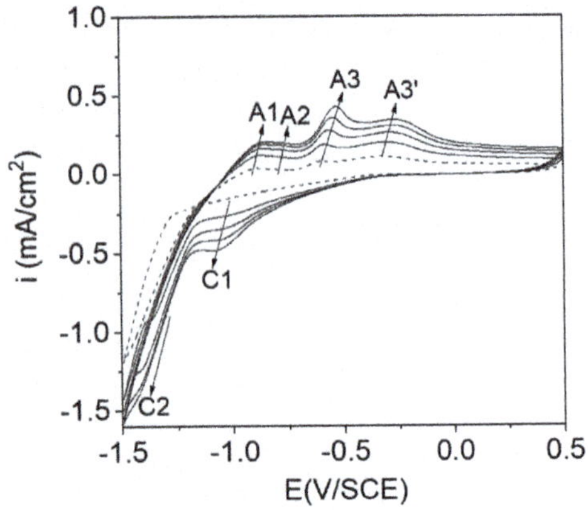

Fig. 6.8 Cyclic voltammetry curves of mill scale

② **Reaction of $Fe^{2+} \rightarrow Fe^{3+}$**: the peak A3 of rebar with mill scale and the descaled rebar has a similar trend with the increase of cycles, so the peak A3 of rebar with mill scale is the $Fe^{2+} \rightarrow Fe^{3+}$ reaction. However, the changing tendency of peak A3' is obviously different between the two rebars. The peak A3' of descaled rebar decreases first and then stabilizes, while that of the rebar with mill scale increases with the cycles, which is more similar to the peak A3' of the mill scale sample, indicating that the reaction at peak A3' is related to the oxidation of the mill scale.

Figure 6.9 is the replotted E-pH diagram at pH = 7–14 of Fe-H_2O system, in which the potential value is referenced to SCE.

It can be seen that in the saturated $Ca(OH)_2$ solution with pH = 12.6, the Fe^{3+} product is more stable. It is considered that peak A3' corresponds to the surface Fe_3O_4 of mill scale oxidizing to Fe^{3+} products.

Obviously, peak C1 is the reverse reaction of the peaks A3 and A3', and peak C2 is the reverse reaction of the peaks A1 and A2.

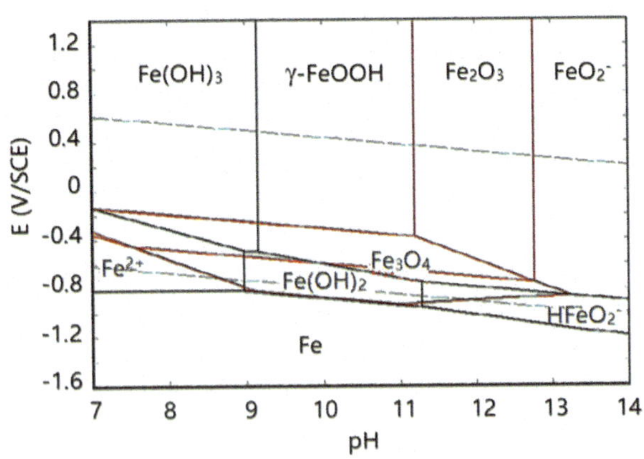

Fig. 6.9 Local E-pH diagram of Fe-H_2O (25 °C, 1 atm) [13]

③ **Reaction of $Fe^{3+} \rightarrow Fe^{2+}$**: there is a reduction peak C1" near −1.2 V in the first cycle of rebar with mill scale, but it does not appear in the first cycle of descaled rebar nor the mill scale sample. Therefore, peak C1" should be related to the reduction of the outer surface of the mill scale. Allanore et al. [15] pointed out that the reduction of $Fe^{3+} \rightarrow Fe^{2+}$ can occur on the surface of hematite (α-Fe_2O_3) at −1.14 V, so it may happen on the surface of rebar with mill scale at C1" peak, and α-Fe_2O_3 can be reduced to Fe_3O_4 or other Fe^{2+} products.

④ **Reaction of $Fe^{2+} \rightarrow Fe^{3+}$ in surface mill scale**: in the "passive" region of −0.05 – 0.5 V, the anodic polarization current of rebar with mill scale and the mill scale sample increases with cycles. Obviously, the latter has no rebar matrix, so there is no traditional passive film produced by the rebar matrix. The anodic polarization current of descaled rebar decreases first and then stabilizes, i.e. there is a passive film formed. This shows that the outer mill scale surface participates in the oxidation reaction in this potential range. Liu et al. [9] believed that Fe_3O_4 in the solid phase can be oxidized in the passive zone. White et al. [16] also pointed out that Fe^{2+} in the solid phase can be oxidized at higher potentials. Therefore, it is considered that $Fe^{2+} \rightarrow Fe^{3+}$ reaction will occur in the surface mill scale (mainly Fe_3O_4) in the potential range of −0.05 – 0.5 V. This has been confirmed by the test results in Chap. 5.

Through the above analysis, the electrode reactions corresponding to the redox peaks of HRB400 rebar with mill scale are listed in Table 6.4.

Table 6.4 Redox peaks of rebar with mill scale and the corresponding electrode reactions

Peak	Potential (V/SCE)	Peak potential moves with cycles	Peak current change with cycles	Electrode reactions
A1	−0.93	→	↑	$Fe^0 \rightarrow Fe^{2+}$
A2	−0.83	→	↑	
A3	−0.58	→	↑	$Fe^{2+} \rightarrow Fe_3O_4$
A3'	−0.36	→	↑	Surface mill scale, $Fe_3O_4 \rightarrow Fe^{3+}$ product
C1	−1.0	←	↑	A3, A3' reverse process, $Fe^{3+} \rightarrow Fe^{2+}$
C1"	−1.2	–	–	α-$Fe_2O_3 \rightarrow Fe_3O_4$ or Fe^{2+} product
C2	−1.3	←	↑	A1, A2 reverse process, $Fe^{2+} \rightarrow Fe^0$
Passive region	−0.05 – 0.5			$Fe_3O_4 \rightarrow Fe^{3+}$ product

6.4 Oxidation and Reduction of Mill Scale

6.4.1 Reduction of Mill Scale

According to Chap. 5, the mill scale on the rebar can be oxidized from outside to inside under anodic polarization and natural immersion in simulated concrete pore solutions. In order to further verify the electrode processes discussed above, the cathodic polarization tests were also carried out.

Figure 6.10 is the Raman spectra along the depths of the HRB400 rebar mill scale section after the rebar cathodically polarized at −1.1 V for 60 days in saturated $Ca(OH)_2$ solution [14]. It can be seen that the inner layer of mill scale is still composed of FeO and Fe_3O_4, but the outer layer is only Fe_3O_4 remained, i.e., the original α-Fe_2O_3 surface layer has disappeared.

Figure 6.11 is the surface Raman spectra of HRB400 rebar with mill scale after cathodic polarization for 7 days and 60 days in saturated $Ca(OH)_2$ solution at −1.1 V. It can be seen that when the rebar is polarized for 7 days, there is still residual α-Fe_2O_3 on the rebar surface at local area (site A), but in another area (site B) α-Fe_2O_3 has completely disappeared. After 60 days of cathodic polarization, only Fe_3O_4 exists on the mill scale surface, the α-Fe_2O_3 has totally disappeared. This is similar to the results observed by Song et al. [3], Sato et al. [17], and Cohen et al. [18, 19].

Figure 6.12 is the cyclic voltammetry curves of HRB400 rebar with mill scale in saturated $Ca(OH)_2$ solution after −1.1 V cathodic polarization for 7 and 60 days. The specimen was immersed in the solution for 12 h before testing.

After cathodic polarization 7 days, peak A3' has a certain negative shift and peak C2 has a positive shift, but it is not significant, indicating that the reaction at peak A3' and peak C2 is more likely to take place after 7 days' cathodic polarization. After cathodic polarization for 60 days, the shape of cyclic voltammetry curves of the rebar has changed significantly, there is one oxidation peak and one reduction peak disappeared, respectively, peak A3' and peak A3 combine into a wide peak, and the corresponding peak C2 and peak C1 do the same. In other words, cathodic

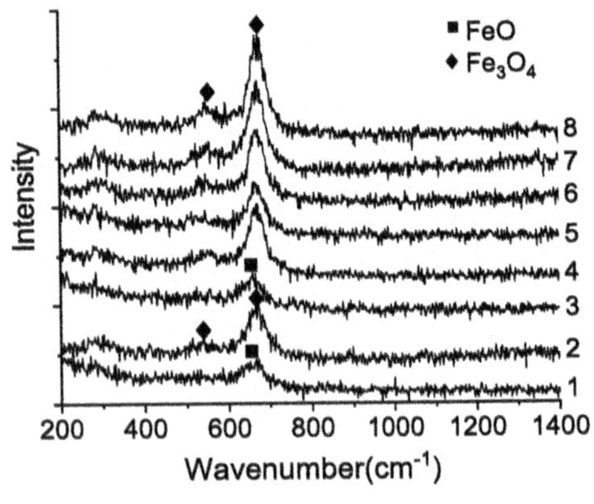

Fig. 6.10 Section Raman spectra of HRB400 rebar after 60d cathodic polarized at −1.1 V. The number 1–8 on the right indicates the depths from the mill scale/matrix to the mill scale/solution interface

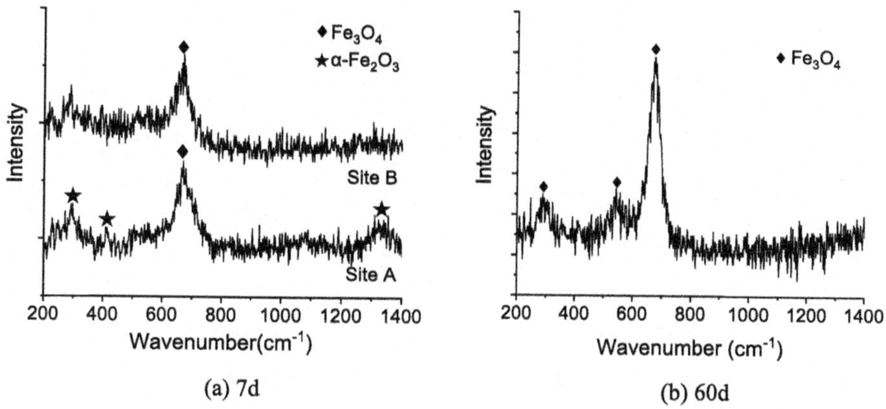

Fig. 6.11 Raman spectrum of mill scale surface of HRB400 rebar with mill scale after cathodic polarization at −1.1 V for different time

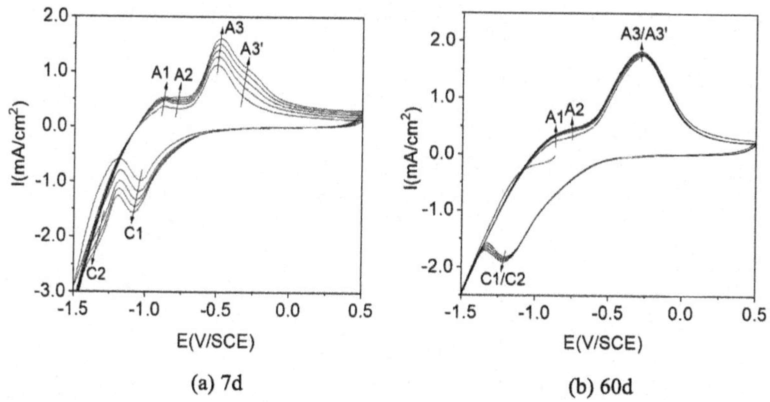

Fig. 6.12 Cyclic voltammetry curves of HRB400 rebar after cathodic polarization for different time at −1.1 V

polarization causes a large amount of Fe^{2+} produced in the mill scale, so the reaction at peak A3/A3' occurs at the mill scale/solution interface and in the surface scale at the same time. The cathodic polarization results in the decrease of Fe^{3+} in the surface layer of the mill scale, so peak C1 negatively moves significantly. The increase of Fe^{2+} in the inner scale makes the reaction of $Fe^{2+} \rightarrow Fe^0$ easier to take place at the scale/matrix interface, i.e., peak C2 moves forward positively, then the opposite movements combine the two peaks into one.

From the above analysis, it can be deduced that the redox reactions of hot-rolled rebar are the redox reactions of the mill scale.

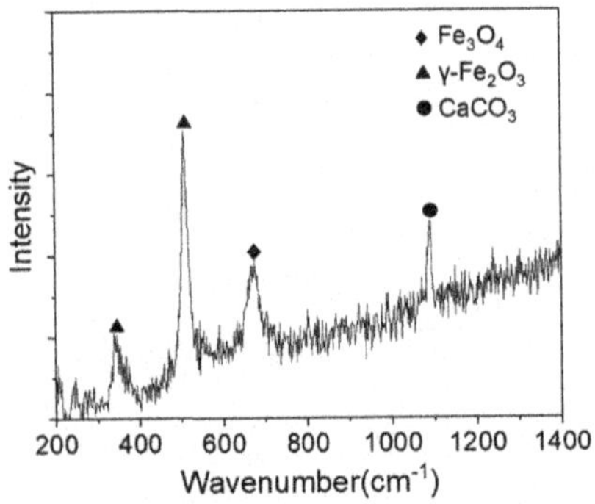

(a) After polarized at 0.3 V for 60d, then polarized at −1.1 V for 7d

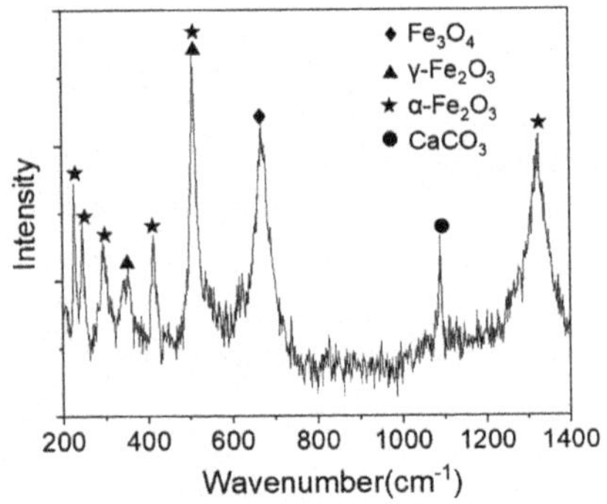

(b) After polarized at −1.1 V for 60d, then polarized at 0.3 V for 7d

Fig. 6.13 Surface Raman spectrum of rebar after reverse polarization

6.4.2 Redox Reaction Mechanisms of Mill Scale

Figure 6.13 is the Raman spectra of HRB400 rebar after reverse polarization for 7 days at −1.1 and 0.3 V, respectively.

According to Fig. 6.13, the anodically pre-passivated mill surface is reduced to Fe_3O_4 and γ-Fe_2O_3, and the cathodically reduced one is reoxidized to α-Fe_2O_3 and γ-Fe_2O_3, which proves again that the oxidation and reduction of the mill scale under applied potentials is "reversible". It also indicates that the Fe^{3+} compounds formed by oxidation on the surface of mill scale can have an α-Fe_2O_3 or γ-Fe_2O_3 form. Since the crystal structures of γ-Fe_2O_3 and Fe_3O_4 are quite similar, γ-Fe_2O_3 should be the intermediate product during the mutual conversion between Fe_3O_4 and α-Fe_2O_3.

Therefore, it can be concluded that the redox reactions of rebar with mill scale in strong alkaline solutions are mainly controlled by the reactions of Fe_3O_4 in the mill scale. During anodic oxidation, with the inward migration of oxygen atoms, Fe^{2+} in Fe_3O_4 on the surface mill scale loses another electron to form Fe^{3+}, and gradually converts to Fe^{3+} oxide. During the cathodic reduction, the opposite process happens, and the Fe^{3+} oxide on the surface mill scale transforms to Fe_3O_4. During strong cathodic polarization, a large amount of Fe^{2+} will be generated in the mill scale, which can be reduced to Fe^0 at the matrix surface. The schematic diagram of the reaction process is shown in Fig. 6.14. It seems that Fe_3O_4 acts as a "buffering solution", which is both a charge carrier and a reaction medium.

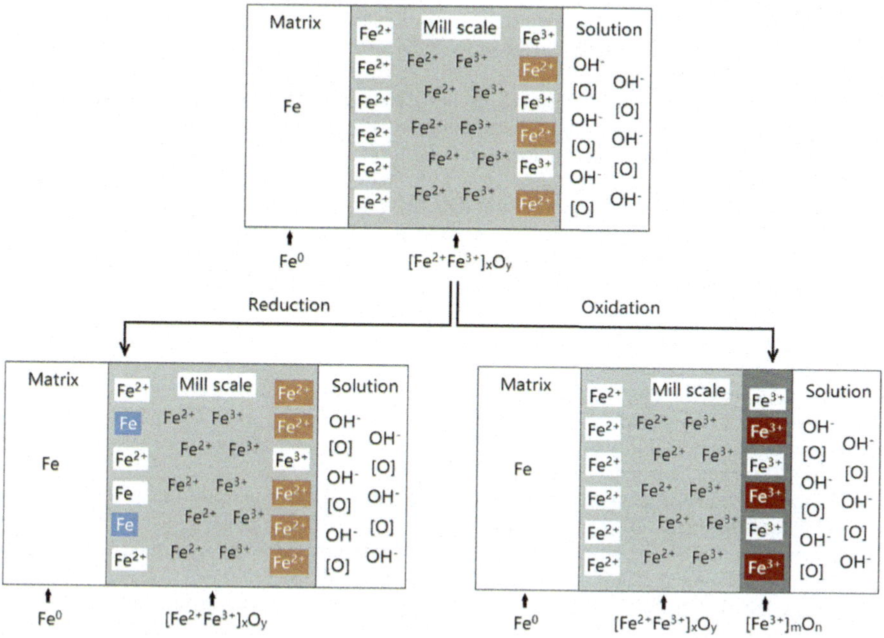

Fig. 6.14 Schematic diagram of oxidation and reduction of the mill scale

6.5 Summary

The reduction of the mill scale of the hot-rolled black rebars is verified through the cathodic polarization, cyclic voltammetry, and Raman spectrum methods. It is found that the redox reactions of black rebars with mill scale in simulated concrete pore solutions are really the redox reactions of the mill scale. The corresponding reaction mechanisms are explained in the present work.

References

1. McCafferty, E.: Introduction to Corrosion Science, pp. 95–118. Springer, New York (2010)
2. Pourbaix, M.: Atlas of Electrochemical Equilibria in Aqueous Solutions, pp. 308–310. Pergamon, Franklin JA Translated. Oxford (1966)
3. Hinatsu, J.T., Graydon, W.F., Foulkes, F.R.: Voltammetric behaviour of iron in cement I. Development of a standard procedure for measuring voltammograms. J Appl. Electrochem. **19**, 868–876 (1989)
4. Song, I., Gervasio, D., Payer, J.H.: Electrochemical behaviour of iron and iron oxide thin films in alkaline (1 M KOH) aqueous solution: a voltammetry study for cathodic instability of coating/metal interface. J Appl. Electrochem. **26**, 1045–1052 (1996)
5. Volpi, E., Olietti, A., Stefanoni, M., et al.: Electrochemical characterization of mild steel in alkaline solutions simulating concrete environment. J. Electroanal. Chem. **736**, 38–46 (2015)

6. Andrade, C., Keddam, M., Nóvoa, X.R., et al.: Electrochemical behaviour of steel rebars in concrete: influence of environmental factors and cement chemistry. Electrochim. Acta **46**(24), 3905–3912 (2001)
7. Sánchez, M., Gregori, J., Alonso, C., et al.: Electrochemical impedance spectroscopy for studying passive layers on steel rebars immersed in alkaline solutions simulating concrete pores. Electrochim. Acta **52**(27), 7634–7641 (2007)
8. Niu, L., Chen, S.H.: Study on the electrode process of iron electrode in alkaline solution. J. Shandong Univ. (Natural science edition) **04**, 494–501 (1990)
9. Mundra, S., Criado, M., Bernal, S.A., et al.: Chloride-induced corrosion of steel rebars in simulated pore solutions of alkali-activated concretes. Cem. Concr. Res. **100**, 385–397 (2017)
10. Xu, W., Daub, K., Zhang, X., et al.: Oxide formation and conversion on carbon steel in mildly basic solutions. Electrochim. Acta **54**(24), 5727–5738 (2009)
11. Harrington, S.P., Devine, T.M.: Relation between the semiconducting properties of a passive film and reduction reaction rates. J. Electrochem. Soc. **156**(4), C154–C159 (2009)
12. Liu, M., Cheng, X., Zhao, G., et al.: Corrosion resistances of passive films on low-Cr steel and carbon steel in simulated concrete pore solution. Surf. Interface Anal. **48**(9), 981–989 (2016)
13. Li, Y.J.: Chemical stability of rebar with mill scale in simulated concrete pore solutions. Master thesis, Tsinghua University (2016). in Chinese
14. Wang, X.: Redox of rebar with mill scale in simulated concrete pore solutions. Master thesis, Tsinghua University (2019), in Chinese
15. Allanore, A., Lavelaine, H., Valentin, G., et al.: Electrodeposition of metal iron from dissolved species in alkaline media. J. Electrochem. Soc. **154**(12), E187 (2007)
16. White, A.F., Peterson, M.L.: The reduction of aqueous metal species on the surfaces of Fe(II)-Containing oxides: The role of surface passivation. In: Sparks, D.L., Grundl, T.J. (eds.). Mineral-water Interfacial Reactions: Kinetics and Mechanisms, 323p. American Chemical Society,Washington (1998)
17. Sato, N., Kudo, K., Noda, T.: Single layer of the passive film on Fe. Corros. Sci. **10**, 785–794 (1970)
18. Cohen, M., Hashimoto, K.: The cathodic reduction of Gamma-FeOOH, Gamma-Fe_2O_3, and oxide films on iron. J. Electrochem. Soc. **121**(1), 42–45 (1974)
19. Oswin, H.G., Cohen, M.: Study of the cathodic reduction of oxide films on iron. J. Electrochem. Soc. **104**(1), 9–16 (1957)

Chapter 7
Corrosion of Hot-Rolled Rebars

Abstract The corrosion mechanisms of rebar with mill scale are discussed in this chapter. The hot-rolled rebars with rust stains are ready to corrode when there is water or aggressive ions, and the rusty rebars are difficult to be re-passivated in concrete. The galvanic corrosion is considered to be the induction of corrosion of the black rebars with mill scale. The main charge carriers are analyzed to be the interstitial metallic ions, not the oxygen vacancies. A unified explanatory model for the passivation and corrosion of rebar is given according to the observations.

7.1 Existing Theories

It is generally believed that the black rebars can be passivated in concrete and form passive films. When the concrete is carbonized or the concentration of corrosive ions (such as chloride) in concrete exceeds the critical value, the rebar's passive film will be damaged and induce the matrix to corrode [1–6]. The volume expansion of the rebar corrosion products can cause the concrete cover cracking and accelerate the rebar corrosion propagation (see Fig. 7.1), and may finally lead the concrete cover spalled.

When chloride ions are free, and the pH value of concrete pore solution is higher than 11.5, the passive film of rebar can remain stable. When the pH value is between 10–11.5, the passive film of rebar is metastable, which is easy to break and cause rebar corrosion. When the pH value is less than 10, the passive film of rebar may disappear and result in large-area corrosion of reinforcement.

If there are chloride ions, even if the pH of the concrete pore solution is no less than 12 and it does not decrease, when $[Cl^-]/[OH^-] \geq 0.6$ [7], the passive film of rebar can break locally and cause a local corrosion.

Table 7.1 lists some critical pH and $[Cl^-]$ values for the corrosion of rebar without mill scale in some publications. It indicates that there are no fixed values for both indexes.

AASHTO-SHRP2-Bridge R19A Appendix C collects some chloride threshold values for carbon steel rebars [12], which are from about 0.02% to 3.08%wt cement, indicating the complication of black rebar corrosion in practice.

X. Lu, *Passivation and Corrosion of Black Rebar with Mill Scale*, Engineering Materials,
https://doi.org/10.1007/978-981-19-8102-9_7

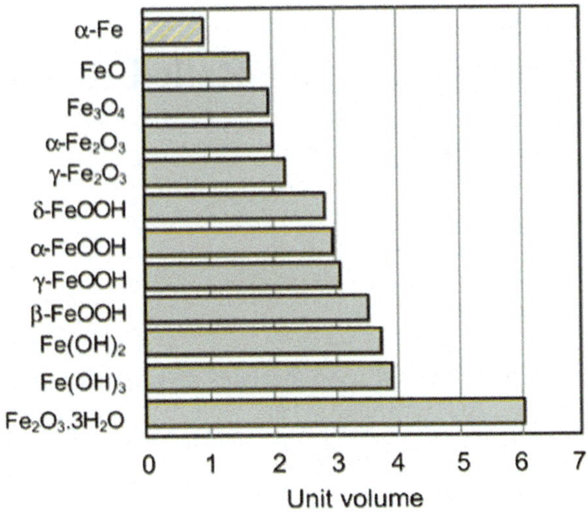

Fig. 7.1 Unit volume of iron and its corrosion products, reprinted from [1] with permission of Elsevier

Table 7.1 Critical pH and [Cl$^-$] of descaled rebars

Solution	pH	Surface state	Methods	Critical pH	Critical [Cl$^-$](M)	Refs.
$Ca(OH)_2$	12.5	1500#	EIS, Raman	–	0.5	Chen et al. [8]
$Ca(OH)_2$	12.5	1500#	XPS	11.31	0.6	Chen et al. [9, 10]
$Ca(OH)_2$	12.5		EIS Polarization curve	11.12–11.05	–	
KOH, NaOH, $Ca(OH)_2$	12.5	polished	Polarization curve	–	0.50–0.55	Liu et al. [11]
	12.0				0.10–0.15	

7.2 Effect of Mill Scale on Metallic Corrosion Resistance

Both metallic corrosion and passivation are surface reactions, so their kinetics is closely related to the surface state of the metals. Therefore, the existence of mill scale will inevitably affect the corrosion resistance of carbon steel.

The effect of mill scale on the corrosion resistance of carbon steel has been controversial for a long time. Up to now, there are three main viewpoints:

(1) **Mill scale can protect the metallic matrix**: some researchers believe that the dense mill scale, especially the one dominated by Fe_3O_4, can improve the corrosion resistance of the metallic matrix [13–15]. Cao et al. [16] have conducted a detailed analysis on the oxide scale' structure which is resistant to atmospheric corrosion, and believe that the mill scale dominated by magnetite and embedded with a sawtooth shape in the matrix has better corrosion resistance. It is also

one of the reasons to improve the atmospheric corrosion resistance of carbon steel by the controlled rolling and cooling processes.

(2) **Mill scale only has physical isolation effect**: the mill scale plays a role in physical isolation for aggressive media from the matrix, there is not much difference after the corrosive media has reached the matrix by penetration through the pores, cracks, or any other micro defects in the mill scale [17, 18].

(3) **Mill scale has negative effects on the corrosion resistance of metallic matrix**: Zhang et al. [19, 20] believe that the discontinuity of oxide scale is an important cause for the local corrosion of the metallic matrix. In addition, the oxide scale can also form a galvanic couple with the matrix and result in serious corrosion. Han et al. [21] believe that the mill scale can promote metallic corrosion.

So far, the research on the corrosion and passivation behavior of mill scale on black rebars is not systematic nor in-depth. Most of the existing studies accept that the mill scale has a negative effect on the corrosion resistance of rebars. For example:

(1) **Mill scale will affect the formation of passive film**: Gouda and Mourad [22] believe that the existence of mill scale will affect the formation and the quality of the passive film on the surface of rebar. When the mill scale forms a galvanic couple with the matrix, it will promote the corrosion of the rebar matrix which cannot be passivated easily.

(2) **Mill scale will prolong the passivation of rebar**: Poursaee et al. [23] consider that the mill scale will prolong the passivation time of rebars, which may need at least 3 days. Ghods et al. [24] think that the hot-rolled rebar in simulated concrete pore solutions will not be stable in the first 8 days.

(3) **Mill scale will reduce the critical chloride concentration**: the test of critical [Cl^-] under different surface states of rebars by Mohammed and Hamada [25] showed that the order of critical [Cl^-] value from large to small was: pre-passivated rebar > rebar without mill scale>rebar with yellow rust>rebar with mill scale. Mammoliti et al. [26] tested the corrosion rate of three kinds of rebars with different surface roughness (including bright and with mill scale) in simulated concrete pore solutions with pH=13.0. They considered that the surface roughness can reduce the critical [Cl^-] value. Pillai et al. [27], Li and Sagues [28] also found that the critical [Cl^-] value of bright steel bars is higher. Xing et al. [29] and Chen et al. [30] studied the corrosion behavior of hot-rolled, rusted, and bright rebars. They considered that the passivation performance of hot-rolled rebars decreases with the decrease of pH and the increase of [Cl^-]. The research of Lin et al. [31] showed that the hot-rolled rebar is more prone to corrosion when the concrete has Cl^- invasion. Chen et al. [32] studied the corrosion behavior of three different grades of hot-rolled rebars in simulated concrete pore solution, and found that when pH=12.5, [Cl^-]=0.05 M can cause the passive film broken and a serious pitting corrosion.

(4) **Mill scale will reduce the repairability of passive film**: Mahallati and Saremi [33] think that the existence of mill scale can reduce the polarization resistance and the repairability of the passive film.

(5) **Mill scale will cause crevice corrosion**: Ghods et al. [34–36] consider that the crevice between the mill scale and the substrate provides favorable conditions for the occurrence of local corrosion, and crevice corrosion is the main cause of the corrosion of rebars with mill scale.

From the above points of view, the corrosion, passivation, and re-passivation behaviors of hot-rolled rebars are still worthy of investigations in-depth, as some essential aspects have not been clarified yet. For example, can the rusted rebars in the atmosphere be re-passivated after concrete pouring? Is there a passive film on the actual hot-rolled rebars with mill scale in concretes? etc.

7.3 Corrosion Resistance of Rusty Rebar

7.3.1 Rusty Rebar

Generally, the surface of the reinforcement just delivered from the steel factory is free of rust and bright gray, as shown in Fig. 7.2a. After exposure to the air for a short period, rust stains will appear on the surface, as shown in Fig. 7.2b. In case of rain, orange-yellow rust may appear overnight, as shown in Fig. 7.2c. If the rebar is placed in the air for a long time, it will be completely and seriously rusted, as shown in Fig. 7.2d.

According to the construction code, the rusted rebars as shown in Fig. 7.2c can be used only after de-rusting, while the seriously rusted rebars as shown in Fig. 7.2d cannot be used in the structure. However, rusty reinforcements are common in field, as shown in Fig. 7.2e and f. Their corrosion resistance may be questionable. However, for the case with minor stains as shown in Fig. 7.2b, is it no problem really?

The corrosion of Fe or low carbon steel in atmospheres, such as rural, urban, industrial, or marine atmospheres, is much complex [37–41]. *The process can be divided into three stages, the first is the formation of a 1–4 nm thick oxide or hydroxide film; the second is the formation of green rust; and the third is the transformation of green rust into a fragile and brown iron oxide or hydroxide.*

The reaction duration of the first stage is usually in the order of milliseconds to seconds, while the second stage can occur in 2–3 h. The production of γ*-FeOOH is a sign of the beginning of the third stage, which is generally observed in 2 weeks or less. A few days after the third stage, it will generate* α*-FeOOH and* Fe_3O_4.

When any anions in the environment participate in the reaction, the diffusion of the oxidizing substances in the rust layer is usually the rate-controlling step. When the corrosion enters the third stage, the corrosion kinetics of iron or ordinary carbon steel usually presents an approximate linear law. The corrosion rates in rural, urban, industrial, and marine atmosphere are about 4–65 μ*m/y, 23–71* μ*m/y, 26–175* μ*m/y, and 26–104* μ*m/y, respectively* [37]. *The corrosion rate of weathering steel is much lower than that of ordinary carbon steel. Its average corrosion rate in the urban atmosphere is about 4–10* μ*m/y.*

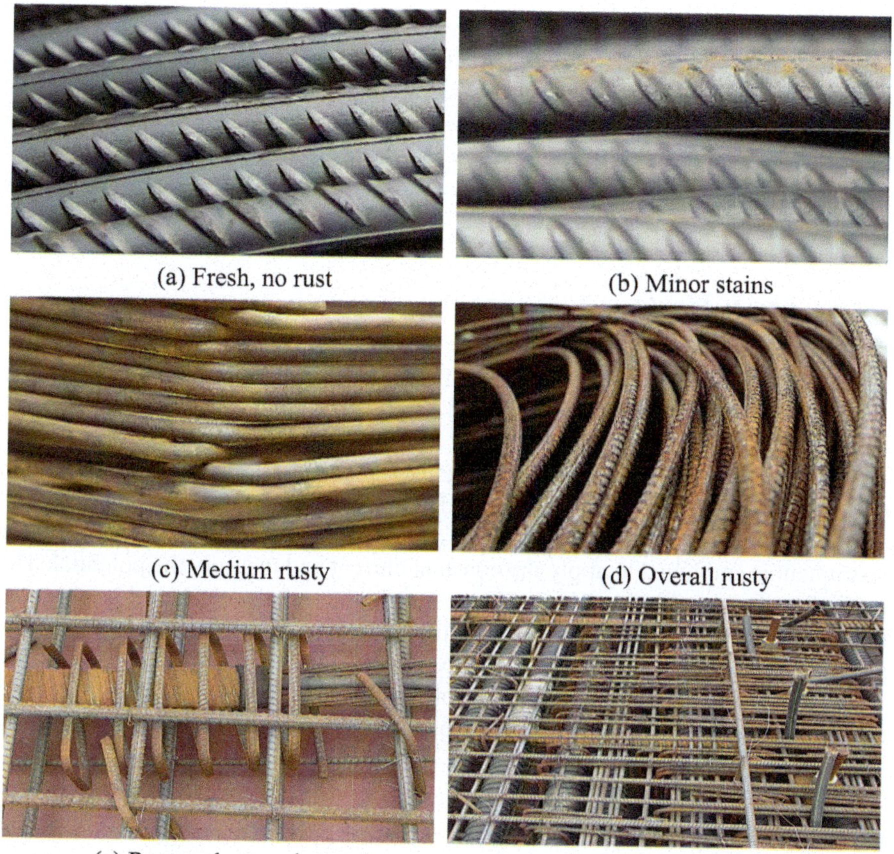

(a) Fresh, no rust (b) Minor stains

(c) Medium rusty (d) Overall rusty

(e) Rusty rebars and strands (f) Rusty rebars on-site

Fig. 7.2 Rusty reinforcements

It should be reminded that the corrosion product of FeOOH detected in the rust should not be confused with the passive film. Please never regard it as the evidence of passive film.

When Fe is exposed to acid moisture [38], *such as acid rain with a pH = 3–6, it is oxidized to* Fe^{2+} *by oxygen or water first, then the green rust forms soon. Some anions from the environment may be embedded in, such as carbonate, sulfate, or chloride. Further oxidation of the green rust may produce FeOOH, which is the most common corrosion product.*

Generally, the inner layer of the rust is mostly a dense amorphous FeOOH with a small amount of Fe_3O_4*, and the outer layer is loose* α*-FeOOH,* γ*-FeOOH, or* γ*-*Fe_2O_3*. When there is an appropriate amount of chloride ion in the environment, it may form* β*-FeOOH, and further be oxidized to* δ*-FeOOH.*

The rust composition of weathering steel containing Cu is slightly different. The inner layer is α*-FeOOH or* γ*-FeOOH with stronger adhesion to the substrate, and the outer is composed of amorphous hydroxide or* δ*-FeOOH.*

The corrosion of Fe in marine environment is not only affected by anions such as chloride and sulfate, but also affected by the concentration of dissolved oxygen in seawater, marine organisms, sulfate bacteria, or other microorganisms [42, 43].

7.3.2 Fresh Mortar

Figure 7.3 is the test results of Φ8mm HPB235 rebar polished, with mill scale and with minor rust stains (labelled as "Rusted") in fresh mortar [44] according to the test method in Appendix B of GB8076-1997, which is a galvanic polarization method. The mortar is prepared by standard cement and sand. Both the water/cement ratio and the cement/sand ratio are 0.5. Distilled water is used as the mixing water. The pH value measured by pH-2s acidity meter is 13.1. PS-6 type reinforcement corrosion instrument is employed to apply the constant current and measure the potential. The applied constant current density is $50\mu A/cm^2$. In the test, the machined bright rebar is used as auxiliary electrode and the saturated calomel electrode (SCE) is used as reference electrode. Before testing, the specimens stand for 5 min (Fig. 7.5).

It can be seen from Fig. 7.3 that the polished rebar and the fresh rebars with mill scale demonstrate a passive characteristic, while the one with mill scale and minor rust stains is in an active state.

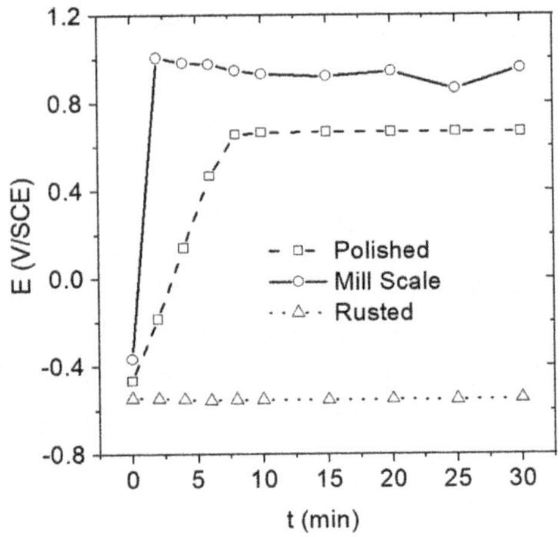

Fig. 7.3 Constant current polarized results in fresh mortar

7.3.3 Cement Hydration Solution

Figures 7.4, 7.5 and 7.6 are the results of galvanic polarization results of the three kinds of rebars mentioned above in cement hydration solutions with or without chloride ions [44]. The cement hydration solution is prepared with an appropriate amount of cement and distilled water, precipitated, and clarified after magnetic stirring for 24 h, with pH≈13.0, in which [Cl^-] is adjusted by adding analytical pure NaCl.

The auxiliary electrode used in the test was a titanium plate, the reference electrode was SCE, and a salt bridge of self-made Lukin capillary filled with 2% agar saturated KCl was used. PS-168A electrochemical testing system was employed to apply the constant current and automatically record the potential-time curves.

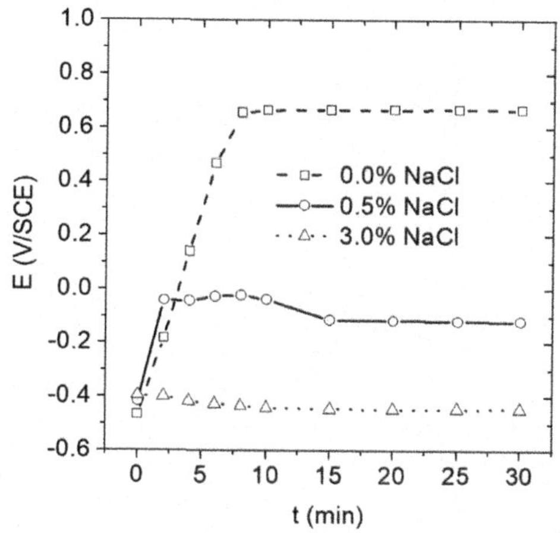

Fig. 7.4 Anodic polarization results of polished rebar

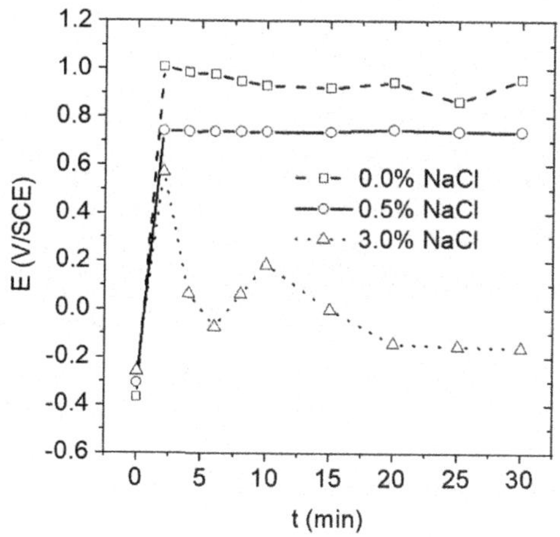

Fig. 7.5 Anodic polarization results of fresh rebar with mill scale

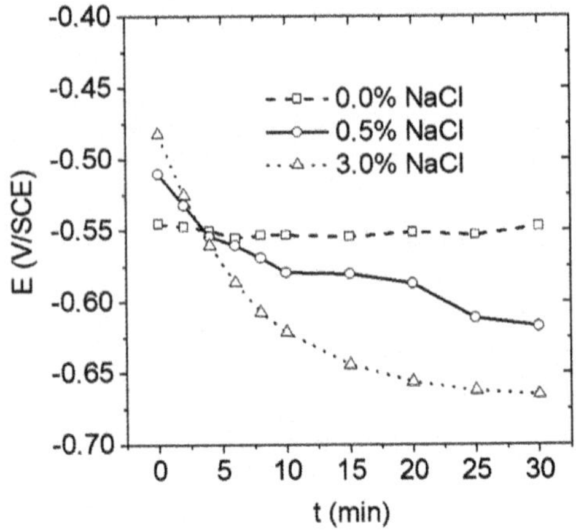

Fig. 7.6 Anodic polarization results of rebar with mill scale and minor rust stains

When there is no chloride salt, the polished rebar and the fresh rebar with mill scale can be passivated, and the anodic polarized potential is continuously stable above 0.6 V.

When the solution contains 0.5% NaCl, only the fresh rebar with mill scale can be passivated with the anodic polarized potential stable above 0.6 V. While for the polished rebar, its potential rises first but decreases rapidly, and stable at about −0.1 V, which is indicating the rebar is unable at such condition.

When the solution contains 3% NaCl, the three kinds of reinforcements are all in active states. The potential of the fresh rebar with mill scale first rises and immediately decreases, and finally stabilizes at about −0.15 V, while the potential of polished one continues to be at about −0.4 V.

It should be noted that the rebar with minor rust stains is in the active state all the time whatever the solution with or without chloride ions, see Fig. 7.6.

7.3.4 *Simulated Concrete Pore Solution*

Table 7.2 lists the self-corrosion currents of the three kinds of rebars mentioned above in chloride-free, simulated concrete pore solutions with pH = 10.0–13.5 [44] (testing conditions same as Fig. 3.1).

It can be seen from Table 7.2 that:

(1) the corrosion rate of rebar with mill scale and minor rust stains is one or two orders of magnitude higher than that of the polished one at pH = 12.5;
(2) when pH < 12.5, the mill scale on fresh rebar seems to have a certain protective effect (initial isolation effect); but when pH ≥ 12.5, there is a slight negative effect, while when pH =13.5, this negative effect becomes obvious since the iron oxide becomes unstable under such a condition.

Table 7.2 Self-corrosion current density of rebars at different pH (μA/cm^2)

Specimen\sol	pH = 10.0	pH = 11.5	pH = 12.5	pH = 13.5
Polished	1.5(94)	0.44(2.8)	0.16(1.0)	0.27(1.7)
Mill scale	0.084(0.53)	0.087(0.54)	0.24(1.5)	0.73(4.6)
Rusted	52(331)	46(288)	7.1(44)	10(63)

Note The values in brackets are folds of the test result of the polished rebar at pH = 12.5

Table 7.3 Self-corrosion current density of reinforcement under different [Cl^-] (μA/cm^2)

Specimen\Sol	pH = 12.5, 0%NaCl	pH = 12.5, 1.0%NaCl
Polished	0.075(1.0)	0.31(4.1)
Mill scale	0.019(0.25)	0.27(3.6)
Rusted	24(320)	118(1573)

Note The values in brackets are folds to the test result of the chloride–free, polished rebar

Table 7.3 is the comparison of self-corrosion currents in simulated concrete pore solutions with and without 1.0%wt NaCl at pH = 12.5 [36]. The corrosion rate of rebar with mill scale and minor rust stains is two or three orders of magnitude higher than that of the polished one without chloride at pH = 12.5. The mill scale without any rust stains has a certain protective effect (initial isolation).

Although the corrosion current is calculated according to the geometric area of the rebar surface, it is absolutely certain that the rebar with mill scale and minor rust stains can significantly corrode in concrete pore solutions with or without chloride ions. This was also verified by Novak et al. [45]. Therefore, never use the results obtained from the polished rebars to infer the corrosion resistance of hot-rolled rebars with mill scale, let alone the rusted rebars!

It can also be seen from Fig. 7.3 that in the fresh mortar, even a 50μA/cm^2 anodic current density cannot force the slightly rusted rebar to passivate. Therefore, the pre-rusted rebar embedded in concrete has a great corrosion risk. As long as there is water or any corrosive medium, it would start to corrode. It needs a special attention in practice.

Therefore, for actual projects, any corrosion of the rebar shall be avoided before concrete pouring. Then, how to improve the corrosion resistance of hot-rolled rebars in atmosphere has become a very important task.

7.4 Corrosion Under Trace Chloride Ions

Figures 7.7, 7.8 and 7.9 are the self-corrosion potentials with time of the polished, descaled, and freshly received HRB335 rebars in saturated $Ca(OH)_2$ + 5 ppm NaCl solution [46]. Here, $[Cl^-]/[OH^-] \approx 1.58 \times 10^{-4}$.

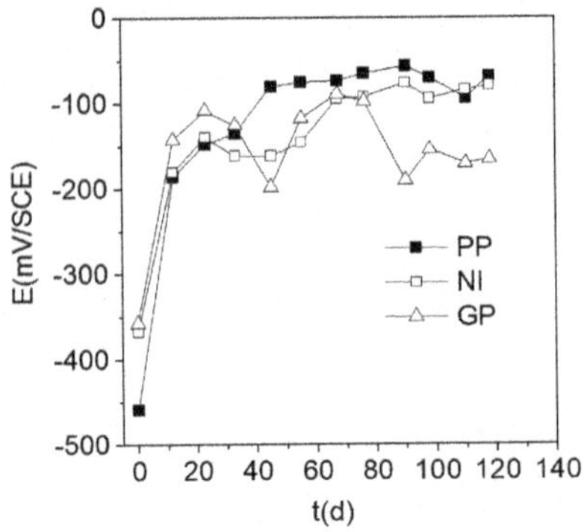

Fig. 7.7 Change of self-corrosion potential of polished rebar

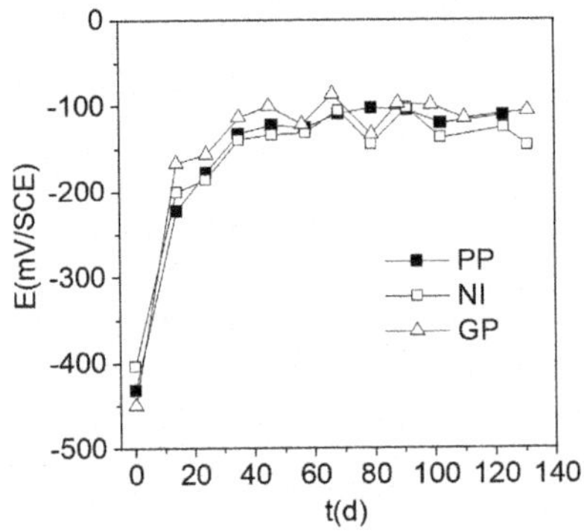

Fig. 7.8 Change of self-corrosion potential of descaled rebar

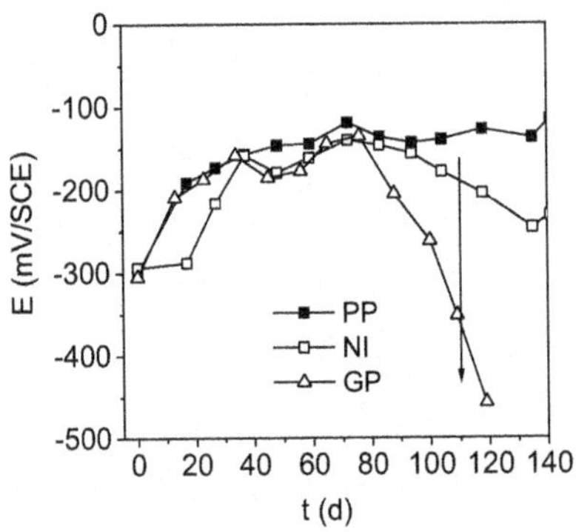

Fig. 7.9 Change of self-corrosion potential of hot-rolled rebar with mill scale

Where, PP in the figures refers to the intermittent potentiostatic passivation at 0.35 V, anodically polarized for 2 h every 10 days; GP refers to the intermittent galvanostatic polarization at 100 μA/cm^2, polarized for 2 h every 10 days, too; and NI refers to the natural immersion in solution. There are 6 parallel samples in each group, and the data in the figures is the average value.

It can be seen that the self-corrosion potentials of polished HRB335 rebar after potentiostatic passivation and natural immersion fluctuate between −50 – −100 mV, while that of the galvanically corroded one changes between −150 – −200 mV. The self-corrosion potentials of the descaled HRB335 rebar are stable at −100 – − 150 mV. However, the self-corrosion potentials of HRB335 rebar with mill scale decrease significantly after natural immersion and galvanic polarization. It indicates that the hot-rolled rebar is much more sensitive to Cl^-.

Figures 7.10, 7.11 and 7.12 are the linear polarization resistances with time of the polished, descaled, and freshly received HRB335 rebars in saturated $Ca(OH)_2$ + 5 ppm NaCl solution.

It can be seen that the linear polarization resistances of the three kinds of rebars increase with time in the early stage of the test duration (0–40d), almost stable in the middle stage (40–80d), while in the later stage (80–140d), the polished and descaled rebars are stable but the rebar with mill scale decreases, and the naturally immersed together with the galvanically polarized ones decrease significantly.

Figure 7.13 is the corrosion morphology and EDS analysis results near the cracks of HRB335 rebar with mill scale after natural immersion in saturated $Ca(OH)_2$ + 5 ppm NaCl solution for 140 days. There are needle-like corrosion products at the cracks. The uplift of the mill scale indicates that it is caused by the expansion of the corrosion products under the mill scale. Chloride is detected in the corrosion product

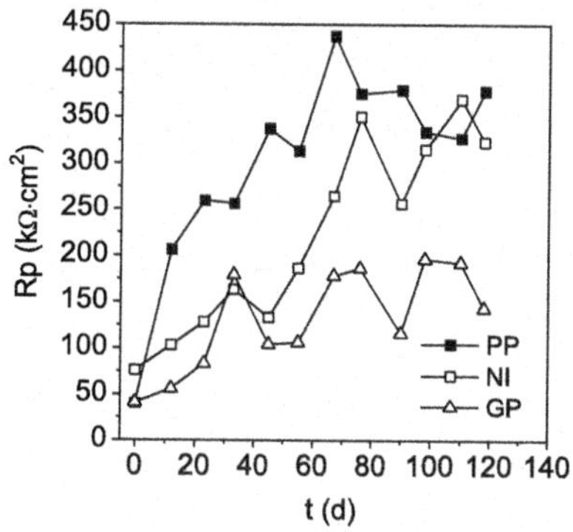

Fig. 7.10 Variation of linear polarization resistance of polished rebar

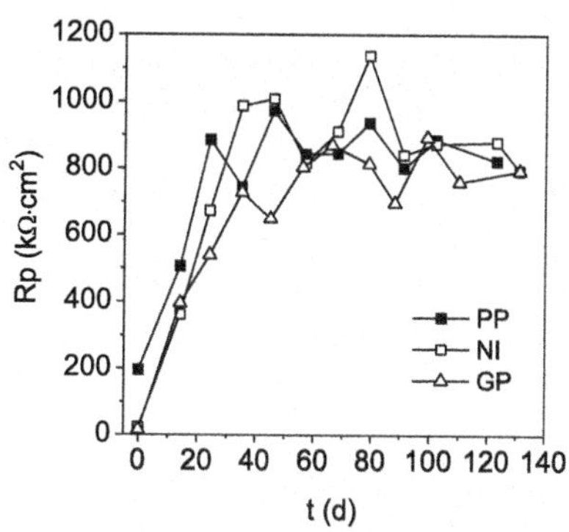

Fig. 7.11 Variation of linear polarization resistance of descaled rebar

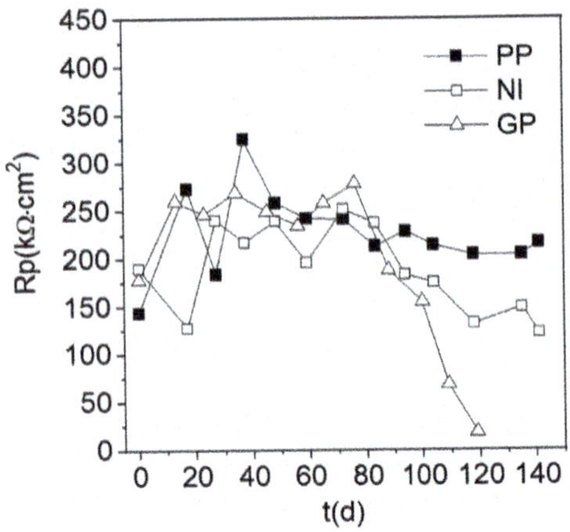

Fig. 7.12 Variation of linear polarization resistance of fresh rebar with mill scale

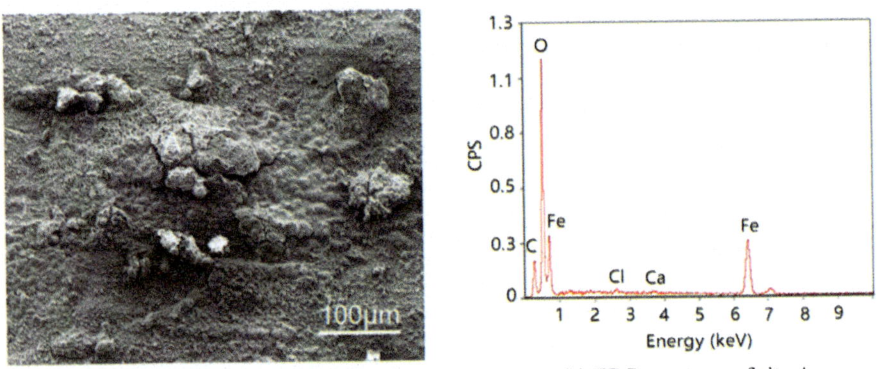

(a) Corrosion morphology (b) EDS spectrum of site A

Fig. 7.13 Corrosion morphology of HRB335 hot-rolled rebar

(site A), but not found in other adjacent areas, which indicates that Cl^- is enriched in the cracks.

Therefore, as long as there is a small number of chloride ions, the corrosion of hot-rolled rebar with mill scale can be induced without requiring a high critical value. The chloride ion concentration in some tap water or groundwater is far more than 5 ppm, so not all the drinkable water is safe for hot-rolled rebars.

7.5 Galvanic Corrosion

From the E-pH diagram (see Fig. 1.6), it can be seen that the iron oxides have higher electrode potentials than iron in aqueous solution. Since Fe_3O_4 is a good electronic

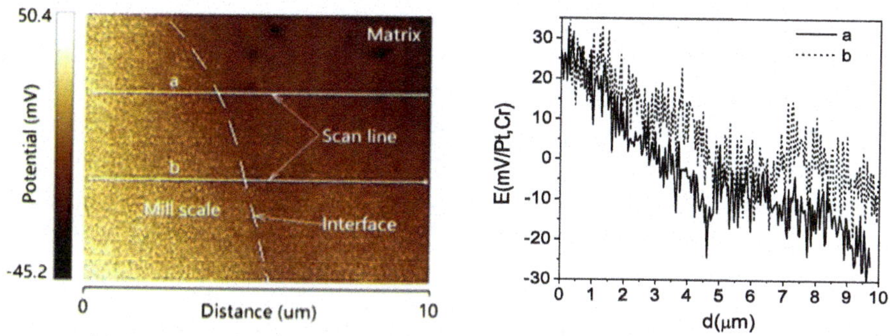

(a) AFM probe potential scanning (b) Linear potential distribution

Fig. 7.14 KPFM image and linear analysis of HRB335 rebar at the mill/matrix interface

conductor, once the mill scale and the substrate contact the aqueous solution or any other electrolytes together, they may form a galvanic couple due to the potential difference, the substrate as anode may turn to corrode in principle.

Figure 7.14 is the potential distribution diagram and the linear potential distribution across the mill scale/matrix interface of HRB335 rebar in atmosphere [46]. It is clear that the potential of mill scale is significantly higher than that of the matrix, the difference is about 55 mV, which is enough for galvanic corrosion to occur.

Since it is difficult to directly measure the potential difference between the mill scale and the rebar matrix in aqueous solution, the HRB335 rebar with mill scale and the descaled one are connected to form a galvanic couple (their surface area ratio is about 5:1) for simulation, by measuring the galvanic current and potential under different pH and Cl^- concentrations, with defining that the galvanic current is positive when the descaled rebar is anodic and negative when the rebar with mill scale is anodic.

Figure 7.15 is the galvanic current of the simulated HRB335 rebar galvanic couple after immersion in different pH solutions without chloride for 24 h. It can be seen that even in different pH solutions without Cl^-, there is still a galvanic current between the hot-rolled rebar and descaled one. When pH $\leq$ 13.0, the galvanic current increases with the decrease of pH, and when pH $\leq$ 12.0, the galvanic current is greater than 4 $\mu A/cm^2$, which can lead to obvious galvanic corrosion. When pH = 13.0, the galvanic current is the smallest. When pH > 3.0, the current increases slightly with pH.

Table 7.4 lists the galvanic current of the simulated HRB335 rebar galvanic couple immersed in saturated $Ca(OH)_2$ solution with different Cl^- concentrations for 1 and 24 h. It can be seen that Cl^- has no significant effect on the galvanic current in saturated $Ca(OH)_2$ solution at the initial stage of immersion. After immersion in saturated $Ca(OH)_2$ solution containing 1000 and 2000 ppm [Cl^-] for 24 h, the galvanic current becomes negative, indicating that there is an electrode polarity reversal between the hot-rolled rebar and the descaled one, and the hot-rolled rebar turns to be anode, indicating that the existence of mill scale increases the corrosion sensitivity of rebar

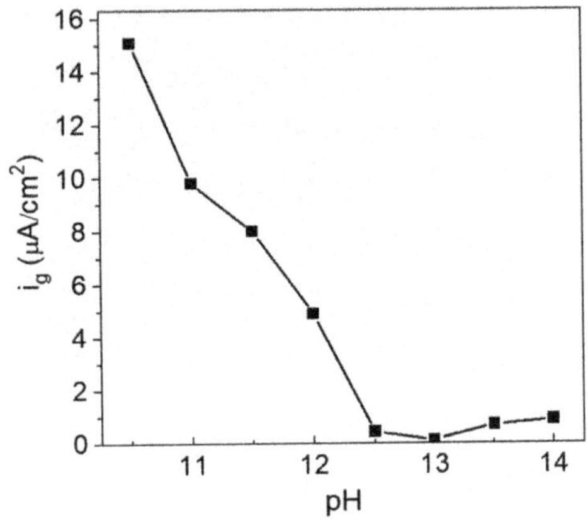

Fig. 7.15 Galvanic current at different pH without chloride

Table 7.4 Galvanic current when immersed in saturated $Ca(OH)_2$ solutions with different $[Cl^-]$ for 1 and 24 h

$[Cl^-]$ (ppm)		0	20	200	1000	2000
I_g(μA/cm²)	t = 1 h	0.77	0.85	0.47	0.22	0.57
	t = 24 h	0.14	0.13	0.15	−0.1	−1.46

when there are enough Cl^- ions. It means that the hot-rolled rebar with mill scale will preferentially corrode at the places where Cl^- can be enriched, such as pores and cracks in the mill scale.

Figure 7.16 is the corrosion morphology of the "matrix-mill scale galvanic couple" of HRB335 rebar (see Fig. C.1 in Appendix C, the apparent surface area ratio S_a: $S_c \approx 1$ between the exposed matrix and the mill scale) immersed in saturated $Ca(OH)_2$ + 1000 ppm NaCl ($[Cl^-]/[OH^-] \approx 3.16 \times 10^{-2}$) solution for 140 days. No corrosion was found in the rebar matrix, but obvious galvanic corrosion occurred at the boundary between the mill scale and the matrix, resulting in needle corrosion products, as shown in Fig. 7.16a. The part with mill scale also has obvious corrosion, and the expansion of corrosion products leads to the damage of the mill scale, as shown in Fig. 7.16b, which indicates that in the presence of trace chloride ions, the galvanic corrosion will preferentially occur at the mill scale/matrix interface where the chloride ions can reach, and then lead to the corrosion of the substrate. It can be inferred that galvanic corrosion is the very first cause of rebar corrosion with mill scale in the presence of aggressive ions.

7.6 Effect of Dissolved Oxygen Concentration

From Chap. 1 it is known that the concentration of dissolved oxygen (abbreviated as DO) in the solution will affect the passivation and corrosion of metals. For the

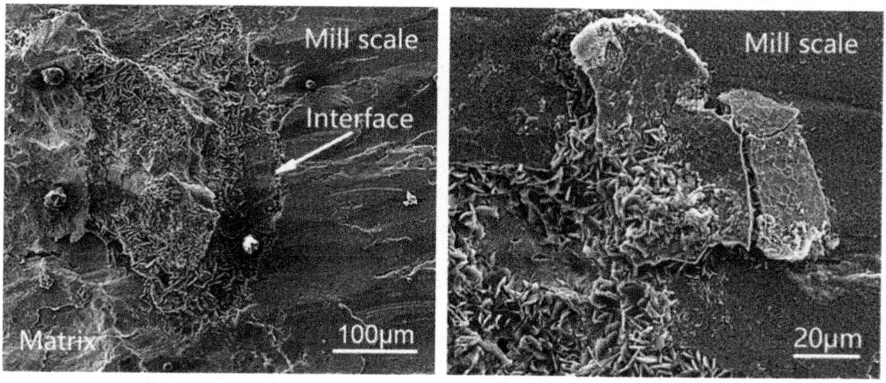

(a) Corrosion at boundary (b) Corrosion at mill scale

Fig. 7.16 Corrosion morphology of the mill scale/matrix galvanic couple of HRB335 rebar

reinforcement in concrete, the dissolved oxygen concentration in the pore solution on the surface of rebar is different under different strength grades of concretes or different thicknesses of concrete covers. Therefore, we use oxygen saturation and nitrogen deaeration methods to study the effect of dissolved oxygen concentration in concrete pore solutions on the rebar's passivation and corrosion, but the synergistic effect with harmful ions such as chloride is not considered temporarily.

Preliminary simulation tests show that the saturated dissolved oxygen concentration in saturated $Ca(OH)_2$ solution is about 9 ppm, and the minimum stable dissolved oxygen concentration obtained by deaeration with high-purity nitrogen is about 0.3 ppm in our lab. Therefore, the passivation and self-passivation of HRB400 reinforcement are investigated under four dissolved oxygen concentrations of 0.3, 2, 4, and 9 ppm. The following test results without indication of dissolved oxygen concentration are all conducted in the open system, in which the dissolved oxygen concentration is about 9 ppm.

Figure 7.17 is the potentiodynamic polarization curves of HRB400 rebars with and without mill scale after immersion in saturated $Ca(OH)_2$ solution for 15 min under different dissolved oxygen concentrations [47].

It is clear that the descaled HRB400 rebar has obvious passivatibility under different dissolved oxygen concentrations, and with the decrease of dissolved oxygen concentration, the self-corrosion potential decreases significantly, while the critical passivation current increases remarkably, and the passive current also increases but slightly (from about 28.5 $\mu A/cm^2$ to about 36.9 $\mu A/cm^2$). The passive potential zone (0.1–0.6 V) remains almost unchanged.

The potentiodynamic polarization curves of HRB400 rebar with mill scale under different dissolved oxygen concentrations are almost unchanged and show typical pseudo-passivation characteristics. The anodic polarization current is small, about

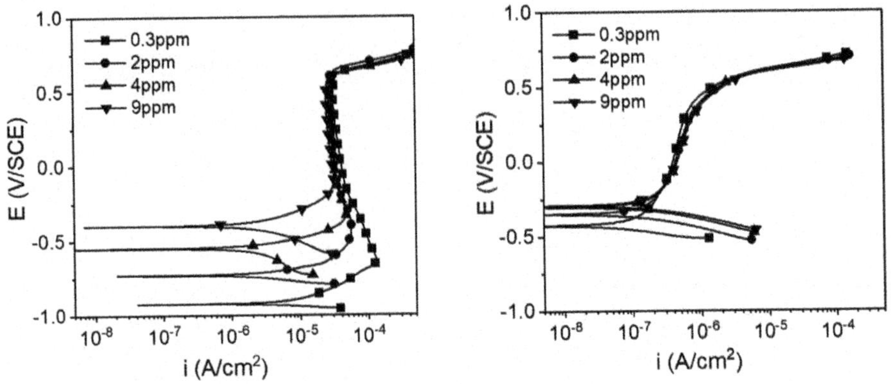

(a) Descaled (b) With mill scale

Fig. 7.17 Potentiodynamic polarization curves of HRB400 rebar under different dissolved oxygen concentrations

0.5 $\mu A/cm^2$. When the dissolved oxygen concentration is less than 2 ppm, the self-corrosion potential of rebar decreases obviously, and the anodic polarization curve at high potentials moves slightly to the left.

Figure 7.18 is the relationship between self-corrosion potential and dissolved oxygen concentration of HRB400 rebar. The higher the dissolved oxygen concentration, the higher the self-corrosion potential, and there is an approximately logarithmic relationship. It is obvious that the dissolved oxygen concentration has a much greater impact on the descaled rebar.

Figure 7.19 is the potentiodynamic polarization curves of HRB400 rebar with and without mill scale after natural immersion in saturated $Ca(OH)_2$ solution with dissolved oxygen concentration of 0.3 and 9 ppm for 15 min and 360 days.

After immersion 360 days, all the polarization curves of the descaled rebar and the one with mill scale move to the upper left, and their shapes are similar. The passive

Fig. 7.18 Relationship between self-corrosion potential and dissolved oxygen concentration of HRB400 rebar

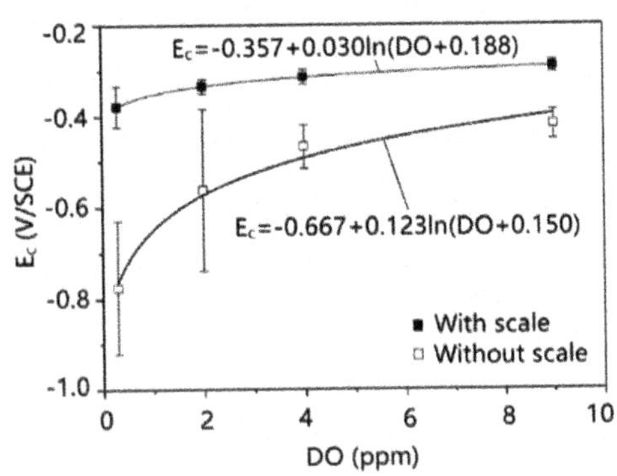

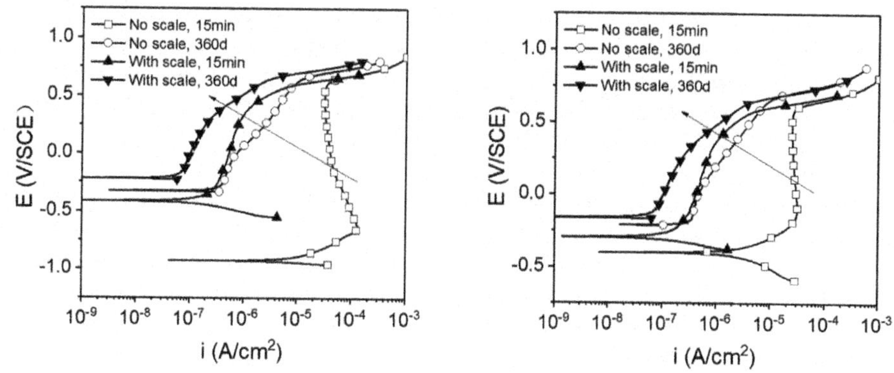

(a) Dissolved oxygen is 0.3ppm

(b) Dissolved oxygen is 9ppm

Fig. 7.19 Potentiodynamic polarization curves of HRB400 rebar after different natural immersion time

current of the descaled rebar is reduced by about two orders of magnitude, and the self-corrosion potential increases by about 0.5 and 0.2 V when the dissolved oxygen concentration is 0.3 and 9 ppm, respectively. The anodic polarization current of the hot-rolled rebar with mill scale decreases to 1/4 – 1/3, and both the self-corrosion potentials increase by about 0.1 V.

After 360 days of natural immersion, the surface of HRB400 reinforcement with and without mill scale has changed, and the shape of its anodic polarization curve is similar. According to the last two chapters, the change of descaled rebar is obviously related to the formation of a passive film, while the change of hot-rolled rebar is related to the oxidation of the surface mill scale.

Figure 7.20 is the self-corrosion potential with time of the HRB400 rebar with and without mill scale in saturated $Ca(OH)_2$ solution under different dissolved oxygen concentrations.

The self-corrosion potential of descaled rebar rises rapidly within the first day after immersion, and the increment increases with the dissolved oxygen concentrations. Except that the change of self-corrosion potential slows down after 3 days at 0.3 ppm, it tends to be stable after the first day at the other dissolved oxygen concentrations. Obviously, this is related to the effect of dissolved oxygen concentration on the growth rate and quality of passive film on the descaled rebar.

Within one day after immersion, the self-corrosion potential of the hot-rolled rebar with mill scale increases slowly under other dissolved oxygen concentrations, except that it decreases first but then increases under 0.3 ppm. After immersion for 7 days, the change of self-corrosion potential tends to be flat, and with the extension of immersion, the self-corrosion potential fluctuates and rises.

Figure 7.21 is the linear polarization resistance with time of the HRB400 rebar with and without mill scale in saturated $Ca(OH)_2$ solution under different dissolved oxygen concentrations.

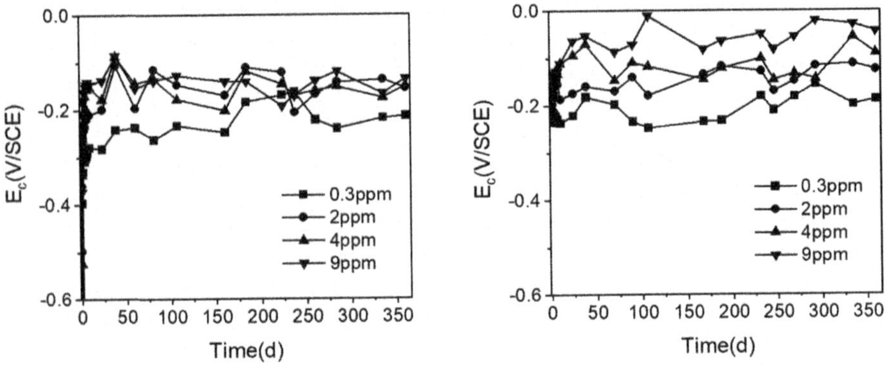

(a) Descaled (b) With mill scale

Fig. 7.20 Change of self-corrosion potential of HRB400 rebar under different dissolved oxygen concentrations

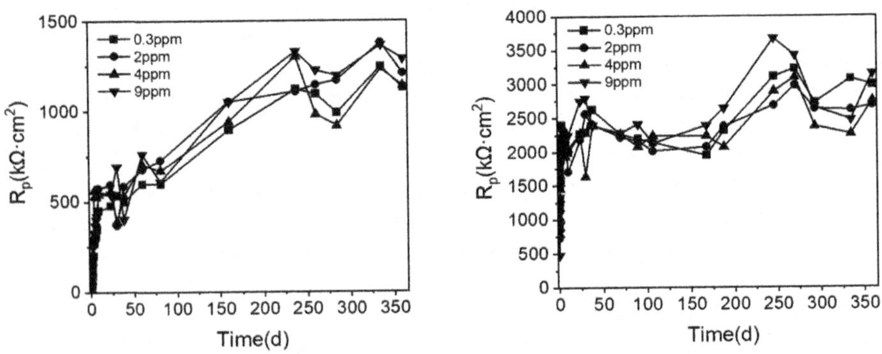

(a) Descaled (b) With mill scale

Fig. 7.21 Linear polarization resistance change of HRB400 rebar under different dissolved oxygen concentrations

It is obvious that the linear polarization resistance of descaled rebar increases significantly with the extension of immersion time, and is positively correlated with the dissolved oxygen concentration, that is, the self-corrosion rate of the descaled rebar decreases significantly with the increase of dissolved oxygen concentration and the passivation time.

The linear polarization resistance of the hot-rolled rebar increases rapidly within the first day of natural immersion, then changes slowly with little increase. Within the testing duration, the dissolved oxygen concentration has no significant effect on the self-corrosion rate of the hot-rolled rebar with mill scale. Comparing the results of both rebars, it is clear that the ohmic drop of the mill scale plays a leading role, that is, physical isolation.

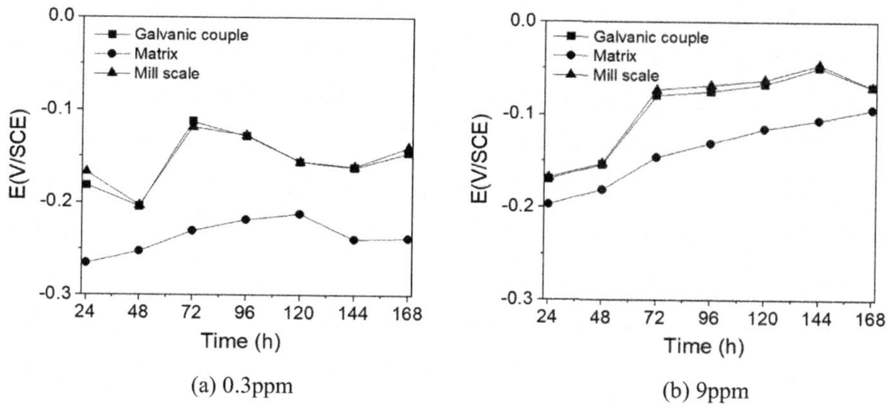

Fig. 7.22 Effect of dissolved oxygen concentration on the rebar matrix, the mill scale, and the galvanic couple potential

In short, the change of dissolved oxygen concentration has a significant impact on the passivation behavior of the descaled rebar, and has an obvious initial effect on the hot-rolled rebars, but the subsequent impact is not as significant as that of on descaled rebar.

Figure 7.22 shows the time-dependent of the HRB400 rebar matrix, the mill scale open circuit potential after the galvanic couple disconnected for 15 min, and the rebar matrix-mill scale galvanic couple potential in saturated $Ca(OH)_2$ solution within 7 days under the dissolved oxygen concentration of 0.3 and 9 ppm.

Under the same conditions, the potential of the galvanic couple is quite close to the open circuit potential of the mill scale, and the potential of the rebar matrix is the lowest, i.e., the rebar matrix is anodic, the mill scale is cathodic, and the electrons flow from the rebar matrix to the mill scale. In addition, the potential of the rebar matrix at the solution concentration of 0.3 ppm is much lower than that of the 9 ppm, and the greater the potential difference with the mill scale, indicating that the rebar matrix becomes more active under such condition.

Combined with the previous test results, it is not difficult to infer that if there are corrosive ions, such as chloride, the rebar matrix is more prone to the preferential corrosion, that is, the hot-rolled rebar matrix embedded in concrete is more prone to corrosion after the penetration of corrosive medium than under the oxygen enriched conditions.

7.7 Carrier Analysis

The formation and destruction of the passive film are related to the transport of the charged particles (i.e., carriers) in the electrochemical reactions. Therefore, the analysis of carriers is helpful to understand the reaction kinetics.

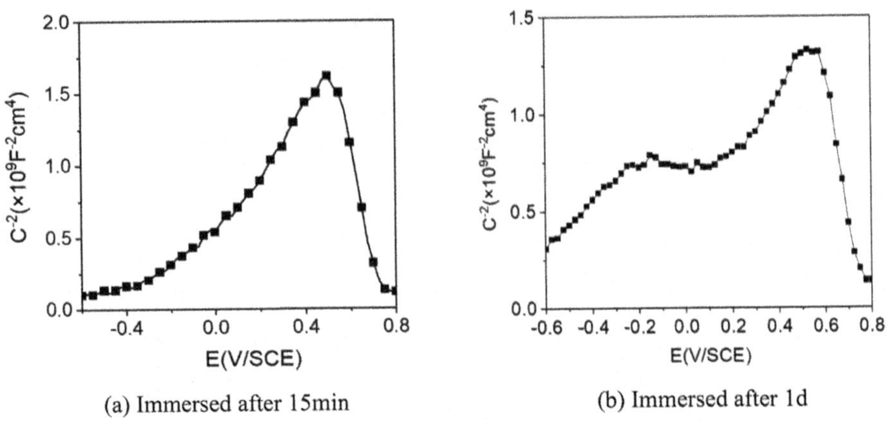

Fig. 7.23 MS curves of descaled HRB400 rebars with different immersion duration

Figure 7.23 is the MS curves of the descaled HRB400 rebars immersed in saturated $Ca(OH)_2$ solution after 15 min and one day [47]. The descaled HRB400 rebar immersed after 15 min has at least two straight line segments with positive slopes in the potential range of −0.4 – −0.5 V, indicating that the electrode has an n-type semiconductor charge transfer characteristic. When the potential is higher than 0.5 V, the slope of the curve becomes negative, showing a p-type semiconductor charge transfer characteristic. At this condition, anodic electrolysis and oxygen evolution take place. Therefore, the slope change of Mott-Schottky curve is closely related to the reactions on the electrode.

Figure 7.24 is the MS curves of HRB400 rebar with mill scale immersed in saturated $Ca(OH)_2$ solution for 15 min and one day [47]. The two curves are similar in shape and have obvious piecewise linear intervals. After immersion for one day, the slope of the line segment increases in the potential range of −0.6 – −0.2 V, but decreases in the potential range of −0.2 – 0.2 V, and changes slightly in the potential range of 0.2 – 0.5 V.

Figure 7.25 shows the overlap of CV and MS curves of HRB400 rebars with and without mill scale after immersion in saturated $Ca(OH)_2$ solution for one day [44]. The potential scanning rate is 25 mV/s. There is an obvious corresponding relationship between the linear segment of MS curve and the oxidation peaks as well as the "passive zone" in CV Curves, i.e., the S1 section corresponds to peak A3, S2 section to peak A3', S3 section to passive region, and S4 section to the transpassive region. According to the electrode reaction analysis in Chap. 6, S1 and S3 are related to the transfer of iron to Fe^{2+} and Fe^{3+}, respectively, while S2 may be related to the mixed transfer to Fe^{2+} and Fe^{3+}.

Figure 7.26 is the MS curves of HRB400 rebar before and after polarization at 0.3 and −1.1 V. After anodic polarization, the slope of segment S1 increases slightly, while segment S2 increases significantly or even disappears, and the slope of segment S3 decreases significantly, which means that the concentration of Fe^{2+} involved in the

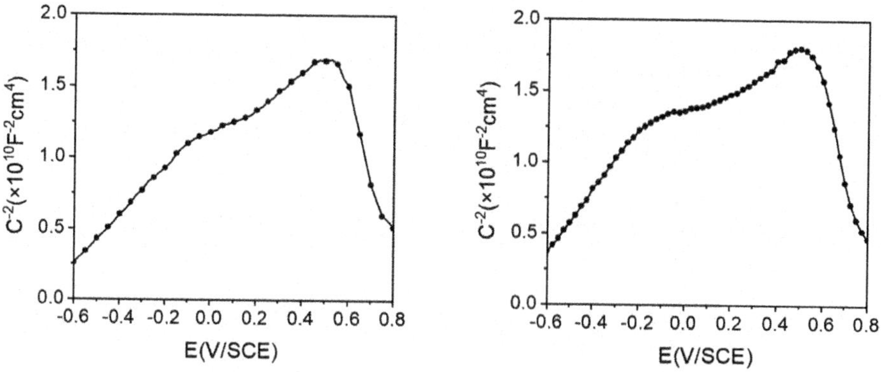

(a) Immersed after 15min (b) Immersed after 1d

Fig. 7.24 MS curves of HRB400 rebar with mill scale after different immersion periods

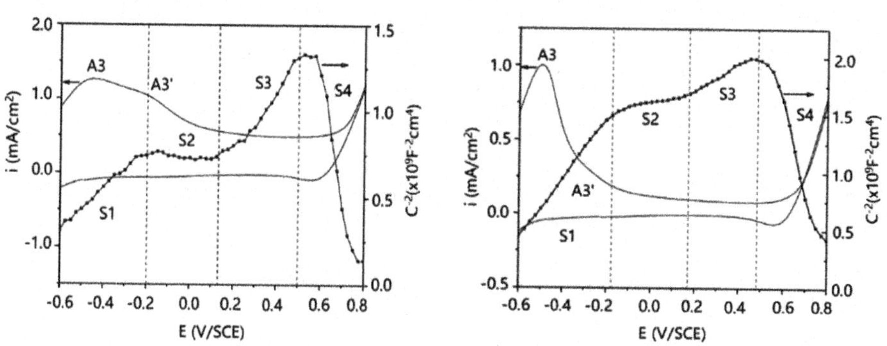

(a) Descaled, immersed after 1d (b) With mill scale, immersed after 1d

Fig. 7.25 CV and MS curves of HRB400 rebar

transfer decreases slightly, while the concentration of Fe^{3+} increases significantly. After the cathodic polarization, it is just the opposite (also see Fig. 7.27), i.e., the slope of S1 and S2 decreases and S3 rises significantly, indicating that the concentration of Fe^{3+} involved in transfer decreases significantly but the concentration of Fe^{2+} increases significantly.

Table 7.5 lists the donor concentrations in different potential regions before and after polarization of HRB400 rebar calculated according to the Eq. (4.1). Since there is no accurate proportion of oxides on the rebar surface, the same relative dielectric constant is taken, $\varepsilon_r = 12$ [48–50], just for comparison temporarily. Anodic polarization is beneficial to improve the carriers in S3 segment, while cathodic polarization is beneficial to improve the carriers in S1 segment. This corresponds to the redox reactions of mill scale in Chap. 6.

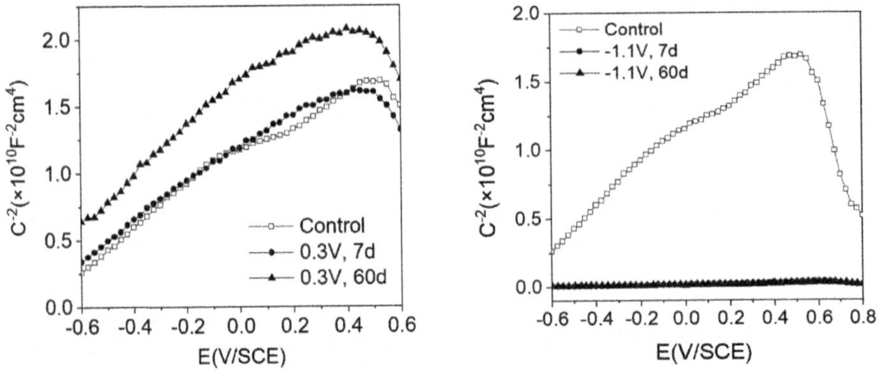

(a) Before and after polarization at 0.3V (b) Before and after polarization at −1.1V

Fig. 7.26 MS curves of HRB400 rebar with mill scale before and after polarization

Fig. 7.27 MS curves of HRB400 rebar after cathodic polarization

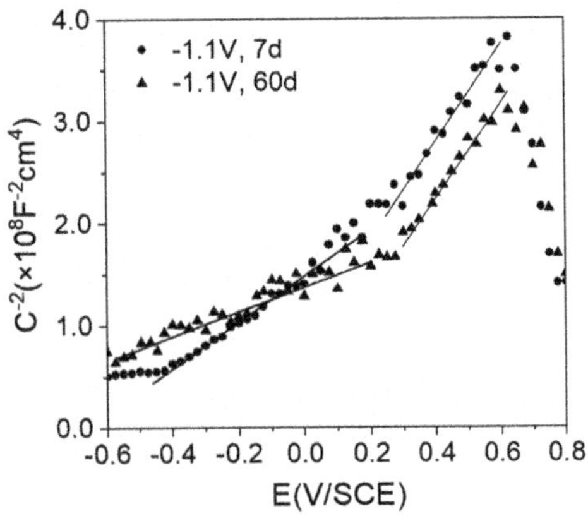

Table 7.5 Donor concentrations of HRB400 rebar before and after polarization (10^{21} cm^{-3})

Fitting potential region	−0.6 – −0.2 V			0.2 – 0.4 V		
Polarization duration	0d	7d	60d	0d	7d	60d
0.3 V anodic polarization	0.679	0.824	0.646	0.822	1.03	1.22
−1.1 V cathodic polarization		48.6	131		24.2	27.6

Traditionally, it is believed that metallic passivation is mostly controlled by the diffusion of oxygen vacancies in the passive film. If this is true, the carrier concentrations of rebar with mill scale should be affected by the oxygen concentrations.

Figure 7.28 is the MS curves of HRB400 rebar with and without mill scale after natural immersion for 360 days in saturated $Ca(OH)_2$ solution at different dissolved oxygen concentrations [47].

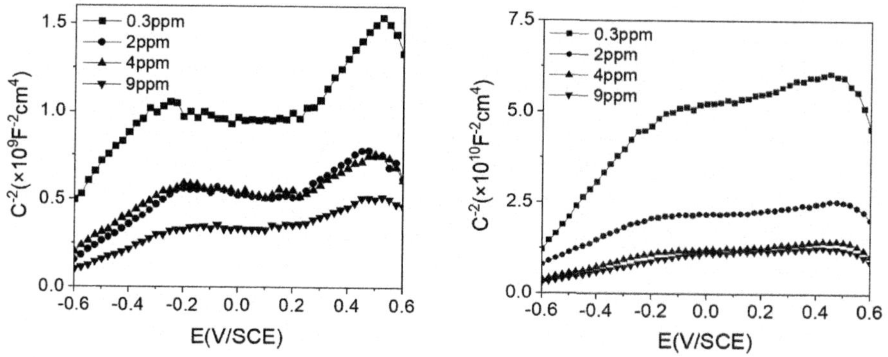

(a) Descaled (b) With mill scale

Fig. 7.28 MS curves of HRB400 rebar after natural immersion for 360d under different dissolved oxygen concentrations

The descaled rebar has three linear segments in the potential range of −0.6 – 0.5 V, with the slope is positive in the range of −0.6 – −0.2 V and 0.2 – 0.5 V, which shows the n-type semiconductor characteristics, and the slope is negative in the range of −0.2 – 0.2 V, which shows the p-type semiconductor characteristics. The higher the dissolved oxygen concentration is, the smaller the slope of segments within −0.6 – −0.2 V and 0.2 – 0.5 V. The case for rebar with mill scale is similar, but the slope of each linear segment is different, which means that the carrier concentration is different.

Table 7.6 lists the carrier concentrations calculated by the MS curve segments after natural immersion for HRB400 rebar with and without mill scale in saturated $Ca(OH)_2$ solution at different dissolved oxygen concentrations for different durations [47]. It shows that:

Table 7.6 Effect of dissolved oxygen on donor concentration of HRB400 rebar (10^{21} cm^{-3})

Surface state	Dissolved oxygen (ppm)	Fitting potential region (V)					
		−0.6 – −0.2			0.2 – 0.4		
		Immersion period (d)					
		1	7	360	1	7	360
Descaled	0.3	19.8	15.8	8.43	8.75	8.71	8.89
	2	18.9	18.0	13.8	8.60	10.0	13.2
	4	15.3	19.0	16.0	5.57	14.9	17.3
	9	10.2	16.8	17.2	5.38	13.5	17.0
With mill scale	0.3	0.748	0.724	0.136	1.41	1.91	0.394
	2	0.826	0.609	0.446	1.55	1.87	1.41
	4	0.763	0.671	0.586	1.38	1.43	1.22
	9	0.647	0.468	0.669	1.20	1.17	1.89

(1) under the same conditions, the carrier concentration in the rebar with mill scale is much lower than that in the descaled rebar, which can explain why the anodic polarization current of the rebar with mill scale is much lower.
(2) from the 360d data, the carrier concentration of the descaled rebar and the rebar with mill scale increases obviously with the increase of dissolved oxygen concentration, which indicates that the carrier in the rebar passive film and the rebar with mill scale cannot be dominated by the oxygen vacancies.
(3) comparing the data at 0.3 and 9 ppm, it is not difficult to find that the influence of dissolved oxygen concentration on the carrier on the descaled rebar is much greater than that on the rebar with mill scale; and according to the data of the descaled rebar, the low dissolved oxygen concentration will reduce the concentration of lower valence carriers in the passive film (i.e., promoting the reduction of passive film), and the high oxygen concentration can increase the concentration of lower and higher valence carriers together (i.e., promoting the growth of passive film).

In general, the carriers of n-type semiconductors are usually oxygen vacancies, interstitial metallic ions, or both. When the main carrier in the material is oxygen vacancy, its concentration will inevitably decrease with the increase of oxygen partial pressure. Obviously, from the above results, whether it is the passive film or the mill scale on the surface of rebar, the main carrier should be iron interstitial ions rather than oxygen vacancies, which is different from the traditional viewpoint. If considering the passive film generated by natural immersion and the main component of the mill scale both are Fe_3O_4, it is not difficult to understand this behavior, because Fe_3O_4 provides rich Fe^{2+} and Fe^{3+} interstitial ions, and these interstitial ions can migrate and convert easily at proper conditions. If the Fe^{3+} concentration is large enough, Fe_3O_4 can transform into a Fe^{3+} compound with the similar crystal structure, for example, the γ-Fe_2O_3.

According to the analysis in Fig. 7.24, the main carrier in the rebar with mill scale should be Fe^{2+} interstitial ions in the potential range of −0.6 – − 0.2 V, and Fe^{3+} interstitial ions in the potential range of 0.2 – 0.5 V.

Therefore, the PDM model [51], which is often used to explain the metallic passivation with the help of oxygen vacancies transfer, may not be suitable for the interpretation of the pseudo-passivation behavior of rebars with mill scale anymore. In fact, Marx [52] also found that it is impossible to use PDM model to better explain the passivation behavior of pure Fe when only consider the oxygen vacancy as the carrier. He suggested that the iron interstitial ions should be considered at the same time.

7.8 Corrosion Mechanisms of Hot-Rolled Rebars

From the above contents, it can be known that the corrosion of the black rebars with mill scale is obviously different from that of the descaled ones. The difference mainly comes from the different characteristics of passive film and the mill scale.

As shown in Fig. 7.29, the descaled rebar can generate a Fe_3O_4 or Fe_3O_4 + γ-Fe_2O_3 passive film with a thickness of several nanometers in strong alkaline and chloride-free environment. But the passive film is not completely ideal, in which there are usually micro defects such as pinholes. The passive film is generally chemically stable in the range of pH=12 $\pm$ 0.5. Beyond this range, the stability of the passive film will decrease. When there are aggressive ions, such as chloride, they can adsorb on the surface of the passive film to form soluble salts, which will dissolve the passive film locally and cause pitting corrosion. They can also preferentially adsorb and enrich at the defects such as pinholes, result in acidification and rapid corrosion at the defect bottom.

Different from the passive film, the mill scale produced by the high-temperature oxidation is about several or tens of microns thick and has micron-scale defects, such as pores, cracks, and even microgrooves. Water and other external corrosive media will preferentially gather in these places due to the capillary suction. Owing to the formation of the galvanic couple between the mill scale and the substrate [22], the matrix will undergo anodic oxidation under strong alkaline conditions without chloride, but the overpotential is limited, it is not enough to passivate the matrix. If the rebar matrix is chemically passivated at the large defects of mill scale, the Fe_3O_4 or Fe_3O_4 + γ-Fe_2O_3 relatively dense passive film is generated. Then the galvanic couple originally formed by mill scale and rebar matrix will reverse the electrodes and gradually go to a stable mixed potential, then the rebar will not corrode. If there are chloride or other active ions, a corrosion couple will be formed between the mill scale and the rebar matrix, so the matrix will be activated until the channel is blocked by the corrosion products, while the corrosion rate may not decrease. At the

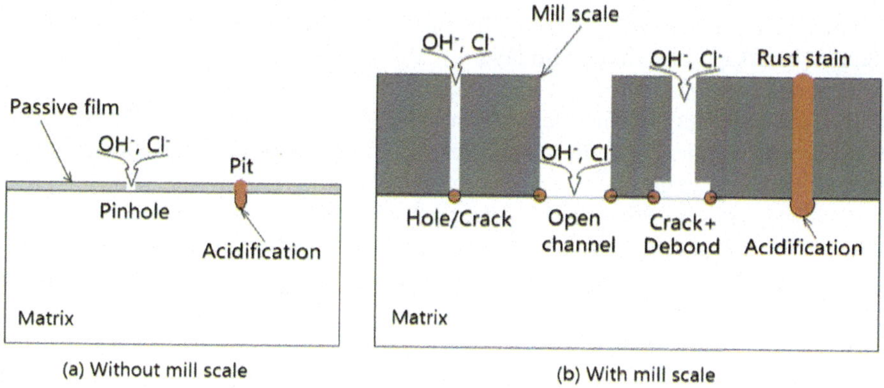

Fig. 7.29 Corrosion diagram of rebar with and without mill scale

debonding places between the mill scale and the rebar matrix, galvanic corrosion occurs first, followed by a rapid crevice corrosion [35]. With the accumulation and expansion of corrosion products, local cracking and spalling of the mill scale will take place, then a rapid corrosion propagation will occur. Actually, there is a big risk of galvanic corrosion and crevice corrosion at any defect of the mill scale. They may take place either sequentially or separately, and the corrosion under the mill scale is usually related to crevice corrosion.

If the corrosion occurs early at the defect of the mill scale, such as those early rust stains, there will be Fe^{2+} or Fe^{3+} ions or their hydroxides appear at the defect bottom, which can cause the local environment to be seriously acidified. If chloride ions or other harmful ions are present, the corrosion at those defects will continue when the conditions are met, see Section 7.3.

The above are some differences in corrosion behaviors between the descaled black rebar and the hot-rolled one with mill scale.

Due to the thickness of passive film is 3 – 4 orders of magnitude less than that of the mill scale, all the passivation or corrosion models involving the high electric field are no longer applicable to the rebars with mill scale. Meanwhile, considering the difference in the main carriers, those passivation and corrosion models dominated by oxygen vacancy diffusion may be no longer applicable to the hot-rolled black rebars with mill scale any more, such as the PDM model [51].

7.9 Unified Model for the Passivation and Corrosion of Rebar

The rebar matrix|passive film or mill scale (film layer)|external medium can be uniformly simplified to the structure as shown in Fig. 7.30, i.e., the film layer is regarded as a solid "buffering" solution of $[Fe^{2+}Fe^{3+}]_xO_y$, which is in the structure of $[Fe^{2+}Fe^{3+}]_xO_y \rightarrow Fe_{1-x}O$ near the interface of rebar matrix, and in the structure of $[Fe^{2+}Fe^{3+}]_xO_y \rightarrow Fe_{2-x}O_3$ near the oxidizing medium, together with the structure of $[Fe^{2+}Fe^{3+}]_xO_y \rightarrow Fe_{3-x}O_4$ in the middle part. Electrons can migrate freely between the Fe^{2+} and Fe^{3+} ions in the film layer.

When the rebar is oxidized, Fe loses electrons at the interface of the rebar matrix |film layer and enters the film in the form of Fe^{2+}, that is, into $[Fe^{2+}Fe^{3+}]_xO_y$. Under the action of chemical potentials or electric fields, Fe^{2+} ions migrate outward through the interstitials. When the reaction at the rebar matrix|film layer is fast, Fe^{2+} may accumulate at this interface, increase the proportion of Fe^{2+}/Fe^{3+} and make it closer to the composition of $Fe_{1-x}O$, which can exist in a large composition range (see Fig. 4.1). While, the oxidants obtain electrons at the interface between the film and the external medium and get reduced. Fe^{2+} at this interface can also be further oxidized to Fe^{3+}, then $Fe_{2-x}O_3$ compounds are gradually formed with the accumulation of Fe^{3+} at this interface, including α-, γ-Fe_2O_3, or any $Fe_{2-x}O_3$ like substance within a certain Fe^{3+}/Fe^{2+} ratio range.

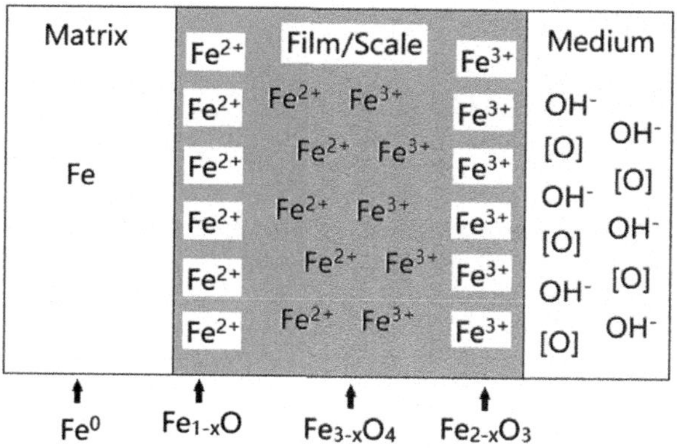

Fig. 7.30 Unified model for passivation and corrosion of rebar

When the rebar is reduced, the reaction process is just the opposite. Fe^{3+} at the interface between the film and the external medium is reduced, while Fe^{2+} at the interface between the rebar matrix and the film layer will accept electrons to be reduced, which may be reduced to Fe under a strong reduction.

If denser the $[Fe^{2+}Fe^{3+}]_xO_y$ film is, the fewer point defects in the lattice, then the greater the impedance, and better the protection of the film, and vice versa. This can explain the differences between the rebar with passive film and the one with mill scale.

According to the above assumptions and the theories that have been verified in the past, all the passivation and corrosion phenomena of rebars with and without mill scale can be explained, including the activated damage caused by harmful ions.

Although Fe_3O_4 should not be regarded as a solid solution of Fe^{2+}, Fe^{3+}, and oxygen nor a mixture of FeO and Fe_2O_3 from the perspective of crystallography, in practice, the above hypothesis can not only explain the passivation and corrosion of reinforcement, but also be extended to the passivation and corrosion of all ordinary carbon steels, which can be used as a unified interpretation model.

7.10 Summary

The differences of corrosion behaviors between the descaled black rebars and the ones with mill scale are comparatively discussed in the present chapter. Galvanic corrosion at the microdefects of mill scale might be the initial corrosion cause of the hot-rolled black rebar in concrete when there are aggressive ions. A unified explanatory model is put forward for the passivation and corrosion of black rebars with and without mill scales according to the experimental observations.

References

1. Poursaee, A.: Corrosion of Steel in Concrete Structures, p. 27. Elsevier, Woodhead (2016)
2. Broomfield, J.P.: Corrosion of Steel in Concrete: Understanding, Investigation and Repair, 2nd edn. Taylor & Francis, London (2007)
3. Raupach, M., Elsener, B., Polder, R., et al.: Corrosion of Reinforcement in Concrete. Woodhead Publishing Limited and CRC Press LLC, Cambridge (2007)
4. Bentur, A., Diamond, S., Berke, N.S.: Steel Corrosion in Concrete: Fundamentals and Civil Engineering Practice. E & FN Spon, London (1997)
5. Mietz, J., Polder, R., Elsener, B.: Corrosion of Reinforcement in Concrete: Corrosion Mechanisms and Corrosion Protection – Papers from EUROCORR '99. European Federation of Corrosion Publications No.31, IOM Communications Ltd, UK (2000)
6. Böhni, H.: Corrosion in reinforced concrete structures. Woodhead Publishing Ltd., Cambridge (2005)
7. Tuutti, K.: Corrosion of steel in concrete. Swedish Cement and Concrete Research Institute, Stockholm (1982)
8. Chen, W., Du, R., Ye, C., et al.: Study on the corrosion behavior of reinforcing steel in simulated concrete pore solutions using in situ raman spectroscopy assisted by electrochemical techniques. Electrochim. Acta **55**(20), 5677–5682 (2010)
9. Chen, W., Du, R., Hu, R., et al.: Correlation between composition of reinforcing steel surface film and steel corrosion behavior in simulated concrete pore solutions. Acta Metall. Sin. **06**, 733–740 (2011). In Chinese
10. Tang, F., Xu, H., Chen, W., et al.: Effect of pH on the electrochemical corrosion behavior of reinforcing steel in simulated concrete pore solutions. Funct. Mater. **02**, 291–293 (2011). In Chinese
11. Liu, Y., Du, R., Lin, C.: Effect of chloride ions on the corrosion behavior of reinforcing steel in simulated concrete pore solutions. Electrochemistry **11**(3), 333–336 (2005). In Chinese
12. AASHTO-SHRP2-Bridge R19A, Service Life Design for Bridges (R19A). https://shrp2.transportation.org/Pages/ServiceLifeDesignforBridges.aspx (2022). Accessed 15 June 2022
13. Gu, R., Wang, H., Song, Y.: Effect of oxide scales on corrosion behavior of Q235 carbon steel. Corros. & Prot. **35**(1), 52–55 (2014). In Chinese
14. Zhou, X., Nie, L., Hua, X., et al.: Influence of heat treatment temperature on structure and corrosion resistance of oxide scales formed on hot-rolled steel 510L. Trans. Mater. Heat Treat. **32**(10), 117–123 (2011). In Chinese
15. Zhu, M.: Influence of phase composition and structure of oxide scales on corrosion resistance of SS400 hot rolled strip. Nanchang aviation university, Master thesis (2011). In Chinese.
16. Cao, Z., Yu, Z., Wang, Y., et al.: The microstructure of scale for corrosion resistant steel. J. Beijing Univ. Iron Steel Technol. **04**, 1–7 (1987). In Chinese
17. He, A.: The effect of oxide film on the electrochemical corrosion behaviors of hot rolled steel. Ocean University of China, Master thesis (2007). In Chinese
18. Deng, J., Hu, J., Deng, P., et al.: Effect of oxide scales on initial corrosion behavior of SPHC hot rolled steel in tropical marine atmosphere. Chin. J. Chin. Soc. Corros. Protection. **39**(04), 331–337 (2019). In Chinese
19. Zhang, H., Qi, G., Dai, J. et al.: Effect of high temperature oxide scale on Corrosion of substrate steel. Mater. Prot. **06**, 24–25 (1995). In Chinese
20. Zhang, H., Qi, G., Yn, Q.: A Study on galvanic corrosion of scale cinder-carbon steel. Oilfield Surf. Eng. **14**(5), 32–34 (1995). In Chinese
21. Han, J., Yan, H., Huang, Y., et al.: Structural features of oxide scales on weathering steel and their influence on atmospheric corrosion. Acta Metall. Sin. **53**(02), 163–174 (2017). In Chinese
22. Gouda, V.K., Mourad, H.M.: Galvanic cells encountered in the corrosion of steel reinforcement I. Differential pH cells. Corros. Sci. **14**(11–12), 681–690 (1974)
23. Poursaee, A., Hansson, C.M.: Reinforcing steel passivation in mortar and pore solution. Cem. Concr. Res. **37**(7), 1127–1133 (2007)

24. Ghods, P., Isgor, O.B., Mcrae, G., et al.: The effect of concrete pore solution composition on the quality of passive oxide films on black steel reinforcement. Cement Concr. Compos. **31**(1), 2–11 (2009)
25. Mohammed, T.U., Hamada, H.: Corrosion of steel bars in concrete with various steel surface conditions. ACI Mater. J. **103**(4), 233 (2006)
26. Mammoliti, L.T., Brown, L.C., Hansson, C.M., et al.: The influence of surface finish of reinforcing steel and pH of the test solution on the chloride threshold concentration for corrosion initiation in synthetic pore solutions. Cem. Concr. Res. **26**(4), 545–550 (1996)
27. Pillai, R.G., Trejo, D.: Surface condition effects on critical chloride threshold of steel reinforcement. ACI Mater. J. **102**(2), 103 (2005)
28. Li, L., Sagues, A.A.: Chloride corrosion threshold of reinforcing steel in alkaline solutions - open-circuit immersion tests. Corrosion **57**(1), 19–28 (2001)
29. Xing, D., He, J., Wu, B., et al.: Corrosion behaviors of rebars with different surface states. Corros. & Prot. **07**, 325–327 (2006). In Chinese
30. Chen, S., Cao, B., Ma, K.: Effects of pH value and chloride ion concentration on passivation behavior of steel rebar in different surface conditions. Corros. & Prot. **08**, 808–812 (2014). In Chinese
31. Lin, D., Du, Y., Chen, Y., et al.: Corrosion behavior of reinforced steel bar with different surface states in the process of Cl- invasion. Corros. & Prot. **10**, 982–986 (2014). In Chinese
32. Chen, H., Li, H., Chen, P., et al.: Electrochemical corrosion behavior of reinforcing steel in simulated concrete solution. J. Build. Mater. **01**, 131–137 (2013). In Chinese
33. Mahallati, E., Saremi, M.: An assessment on the mill scale effects on the electrochemical characteristics of steel bars in concrete under DC-polarization. Cem. Concr. Res. **36**(7), 1324–1329 (2006)
34. Ghods, P., Isgor, O.B., Mcrae, G.A., et al.: Electrochemical investigation of chloride-induced depassivation of black steel rebar under simulated service conditions. Corros. Sci. **52**(5), 1649–1659 (2010)
35. Ghods, P., Isgor, O.B., Mcrae, G.A., et al.: Microscopic investigation of mill scale and its proposed effect on the variability of chloride-induced depassivation of carbon steel rebar. Corros. Sci. **53**(3), 946–954 (2011)
36. Ghods, P.: Multi-Scale Investigation of the Formation and Breakdown of Passive Films on Carbon Steel Rebar in Concrete. Carleton University (2010)
37. Leygraf, C., Wallinder, I.O., Tidblad, J., et al.: Atmospheric Corrosion. John Wiley & Sons, Inc., New Jersey (2016)
38. Graedel, T.E., Frankenthal, R.P.: Corrosion mechanisms for iron and low alloy steels exposed to the atmosphere. J. Electrochem. Soc. **137**(8), 2385–2394 (1990)
39. Kaesche, H.: Corrosion of Metals: Physicochemical Principles and Current Problems. Springer, Berlin (2003)
40. Legrand, L., Mazerolles, L., Chaussé, A.: The oxidation of carbonate green rust into ferric phases: solid-state reaction or transformation via solution. Geochim. Cosmochim. Acta **68**(17), 3497–3507 (2004)
41. Sagoe-Crentsil, K.K., Glasser, F.P.: "Green rust", iron solubility and the role of chloride in the corrosion of steel at high pH. Cem. Concr. Res. **23**(4), 785–791 (1993)
42. Saha, J.K.: Corrosion of Constructional Steels in Marine and Industrial Environment: Frontier Work in Atmospheric Corrosion. Springer, India (2013)
43. Chandler, K.A.: Marine and Offshore Corrosion. Butterworth-Heinemann, Elsevier Ltd (1985)
44. Lin, W.: Influences on rebar corrosion by rebar surface states, pH of concrete pore solution and [Cl^-] in concrete. Tsinghua University, Master thesis (2002). In Chinese.
45. Novak, P., Mala, R., Joska, L.: Influence of pre-rusting on steel corrosion in concrete. Cem. Concr. Res. **31**(4), 589–593 (2001)
46. Li, Y.J.: Chemical stability of rebar with mill scale in simulated concrete pore solutions. Tsinghua University, Master thesis (2016). In Chinese
47. Wang, X.: Redox of rebar with mill scale in simulated concrete pore solutions. Tsinghua University, Master thesis (2019). In Chinese

48. Hamadou, L., AïnouchE, L., Kadri, A., et al.: Electrochemical impedance spectroscopy study of thermally grown oxides exhibiting constant phase element behaviour. Electrochim. Acta **113**, 99–108 (2013)
49. Wielant, J., Goossens, V., Hausbrand, R., et al.: Electronic properties of thermally formed thin iron oxide films. Electrochim. Acta **52**(27), 7617–7625 (2007)
50. Hakiki, N.E., Montemor, M.F., Ferreira, M.G.S., et al.: Semiconducting properties of thermally grown oxide films on AISI 304 stainless steel. Corros. Sci. **42**(4), 687–702 (2000)
51. Macdonald, D.D.: The history of the point defect model for the passive state: a brief review of film growth aspects. Electrochim. Acta **56**, 1761–1772 (2011)
52. Marx, B.M.: The Mechanisms of Passivity on Iron: Experimental Methods for Characterizing and Developing Models to Describe Nano-Oxide Growth. The Pennsylvania State University, Department of Materials Science and Engineering (2006)

Chapter 8
Protection of Hot-Rolled Rebars

Abstract The principles and measures for protection of hot-rolled rebars are briefly introduced. The high-temperature rapid phosphating to improve the atmospheric corrosion resistance and the cathodic prevention using sacrificial anodes for rebars subject to chlorides are recommended. It should be careful to apply the electrochemical desalination or re-alkalization to the actual rebars with mill scale, and it is suggested to use the ultra-high performance concrete as a repair material to protect the deteriorated normal concrete structures.

8.1 General Principles

The metallic corrosion is such a process in which a metal loses electrons and gets oxidized (see Eq. 1.1). It is a natural phenomenon. People can only delay it or use it, but can't stop it.

The occurrence of any metallic corrosion must have three elements at the same time, namely, the metal, the environment, and the direct interaction between them, see Fig. 8.1. As long as one of them changed, the original corrosion phenomenon would change, shift, or disappear. This is not only a simple approach to find the cause of any metallic corrosion problems, but also the right way to metallic corrosion protections,[1] which is called "3-Element analysis method".

Since the surface state of metals affects their corrosion thermodynamics and kinetics, not all the conclusions on the descaled rebars can be directly applied to the one with mill scale, let alone to the corroded rebars.

Therefore, this chapter only briefly introduces some concerned protective measures, such as the surface modification to anti-corrosion of the black rebar in atmosphere, electron supply to rebars by distributed sacrificial anodes, modification the interfacial environment of rebars using electrochemical techniques and mechanical barriers on the rebar or concrete surface. Any other details are suggested to find elsewhere in the related publications.

[1] Including reasonably selecting materials, improving the environment and changing the interaction path.

X. Lu, *Passivation and Corrosion of Black Rebar with Mill Scale*, Engineering Materials,
https://doi.org/10.1007/978-981-19-8102-9_8

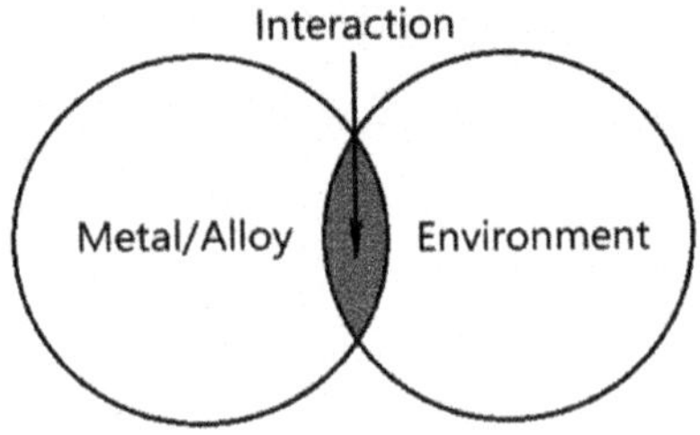

Fig. 8.1 Schematic of the three elements of metallic corrosion

8.2 High-Temperature Phosphating of Hot-Rolled Rebars

It is known from Sect. 7.3 that the rust stains on the black rebar with mill scale formed in the air will become the corrosion starting points at any time when the condition is met. Therefore, ensure the hot-rolled rebar not to rust in the atmosphere before concrete pouring is the first task to be done by the milling factories or by the construction enterprises.

At present, the primary means to improve the atmospheric corrosion resistance of hot-rolled rebars is to convert the mill scale into a relatively dense Fe_3O_4 scale through the controlled rolling and cooling processes [1–3]. Compared with the traditional technology, it can slightly prolong the rusting induction period of the hot-rolled rebar in an ordinary atmosphere. It has little effect on the rebar corrosion after exposure to water, such as rain.

In 2006, Manna et al. [4] from the R & D Department of Tata Steel Company in India published a paper saying that one-step phosphating treatment (without degreasing, cleaning, and pickling) for hot-rolled rebar could effectively improve its corrosion resistance in a humid atmosphere.

They used two kinds of phosphating solutions with and without nitric acid (see Table 8.1) to phosphatize the hot-rolled rebars, and the corrosion resistance of the rebars before and after phosphating was examined by atmospheric exposure, humid cycling, and salt spray tests.

After phosphating, the surface of hot-rolled rebars becomes dense and black in appearance. However, the composition of the phosphating coating obtained by the two phosphating processes is different. The surface of the rebar treated with nitric acid phosphating solution is composed of mainly zinc phosphate (Hopeite)

Table 8.1 Two phosphating solutions with/without nitric acid and phosphating processes, reprinted from [4] with permission of Elsevier

Phosphating system	H_3PO_4 (ml/L)	ZnO (g/L)	HNO_3 (ml/L)	pH	Temp. (°C)	Time (min)
Nitric acid contained	18	10	10	2.00	90	10
Nitric acid-free	28	10	/	2.05	90	10

Table 8.2 Salt spray test results of rebars before and after phosphating treatment, reprinted from [4] with permission of Elsevier

Rebar	Red rust formation (%)			
	5 h	20 h	50 h	100 h
Untreated	10–15	95	100	100
Solution with nitric acid	0	1–2	50	100
Solution without nitric acid	0	0	0	2

and some zinc-iron phosphate (Phosphophyllite), while the surface of the reinforcement treated without nitric acid phosphating solution is mainly composed of zinc phosphate (Hopeite) and some iron phosphate (Ludllamite) [4].

It is found that the hot-rolled rebars without phosphating will produce red rust after exposure to the atmosphere for 2–3 days, while the treated ones by phosphating with and without nitric acid have no red rust within 60 and 180 days, respectively [4]. Therefore, the phosphating treatment can significantly improve the atmospheric corrosion resistance of hot-rolled rebars with mill scale.

Table 8.2 demonstrates the salt spray test results of the hot-rolled rebars before and after phosphating. It indicates that both the phosphating treatments can significantly improve the salt spray corrosion resistance of the reinforcements. The one treated without nitric acid is relatively better [4]. Actually, phosphating treatment is often used for the protection of hot-rolled rebars transported by sea.

In addition, they also conducted the rebar pull-out test in concrete. The results show that the bond strength of hot-rolled rebars treated with the nitric acid-free phosphating increases slightly. However, for the one treated with nitric acid phosphating, the bond strength is decreased by 10–15% [4]. Since that time, they have carried out a series of follow-up studies. Besides phosphating, they have also studied some other processes, such as electroless nickel plating.[2]

Unfortunately, Manna et al. [4] adopted the traditional conventional phosphating method and did not consider using the controlled rolling and cooling processes to directly phosphatize the rebar during or after rolling in the steel plant. Therefore, from 2013 to 2015, we conducted the high-temperature rapid phosphating simulation tests [5–7] for the rebar with mill scale, for example, heated the HRB235 rebar with mill scale at 950 °C for 5 min, then immersed it in phosphating solution for 30 or 60 s. After considering the chemical stability of the phosphating layer in concrete, we selected the zinc-calcium and the zinc-nickel phosphating systems (see Tables 8.3 and 8.4). The phase analysis about the phosphating layer on the rebar was also carried out, and the corrosion resistance was examined by copper sulfate drop test (see Table 8.5) as well as some electrochemical tests, including the corrosion resistance in simulated concrete pore solution and artificial acid rains (see Table 8.6).

It is known from Tables 8.5 and 8.6 that the high-temperature rapid phosphating can significantly improve the corrosion resistance of hot-rolled black rebars with mill scale. This procedure is especially suitable for the rebar with mill scale to be

[2] See https://www.researchgate.net/profile/Manindra_Manna.

Table 8.3 Zn-Ca phosphating solution [5–7]

Component	$Zn(H_2PO_4)_2 \cdot 2H_2O$	$Zn(NO_3)_2$	$Ca(NO_3)_2$	$NaNO_2$	$Ni(NO_3)_2$	$Na_5P_3O_{10}$
Content (g/L)	36	26	32	0.4	3	0.8

Table 8.4 Zn-Ni phosphating solution [5–7]

Component	ZnO	H_3PO_4	EDTA	$Na_5P_3O_{10}$	$NaNO_2$	NaF	$Ni(NO_3)_2$
Content(g/L)	10	60	3	3	5	6	8.4

Table 8.5 Copper sulfate drop test results of rebar before and after phosphating [5–7]

Rebar	Zn-Ca phosphating (s)	Zn-Ni phosphating (s)
Before phosphating	15	15
High-temperature fast phosphating 30 s	40	40
High-temperature fast phosphating 60 s	112	66

treated during the high-temperature rolling or in a very short time just after the final rolling in the steel plant.

It is certain that one of the best ways is to improve the corrosion resistance of hot-rolled rebars from the very beginning in the mill factories. It can not only increase the delivery quality and avoid the premature rust of rebars in storage or transportation, but also improve the durability of reinforced concrete afterwards. Therefore, it is welcome that the civil engineers can clearly label the use of phosphatized rebars in the design documents or in the procurement contracts, to avoid premature corrosion of hot-rolled black rebars before concrete pouring.

Table 8.6 Test result in artificial acid rain solution[a]

Rebar	E_c (V)	i_c ($\mu A/cm^2$)	Time to red rust appears (d)
Untreated rebar	−0.642	172	<1
High-temperature fast phosphating 60 s, Zn-Ca	−0.560	47	~6
High-temperature fast phosphating 60 s, Zn-Ni	−0.423	30	~6

[a] Artificial rain solution: 15 μmol/L NaCl + 25 μmol/L $(NH_4)_2SO_4$, pH = 5 adjusted using HNO_3

8.3 Cathodic Protection

It is known from Sect. 7.3 that the rusty rebar will always be activated under the environment containing chlorides, and it cannot be re-passivated only by the alkalinity of concrete pore solution. For such a case, cathodic protection can be chosen to inhibit the further corrosion of rebar.

According to the different electron sources, cathodic protection can be divided into "impressed current" and "sacrificial anode" types [8], see Fig. 8.2. The former supplies the electrons by a DC equipment, while the latter by the more active (i.e., more sensitive to corrosion) metals in the same environment, such as zinc, aluminum, magnesium alloys (sacrificial anode), etc.

Cathodic protection is a conventional technology in metallic corrosion field. There are some European and American guidelines or standards available for the impressed current cathodic protection, which has been applied to some important metallic structures in petrochemical industry, port engineering, and hydraulic infrastructures [9–11], as well as ships or submarines, etc.

In the field of reinforced concrete, impressed current cathodic protection has been applied to some nearshore or offshore structures, such as some important military and civilian facilities abroad, for more than half a century, and has been gradually used in China in recent 20 years [12–18]. The domestic engineering application of sacrificial anode cathodic protection is relatively few. At present, it is mainly used in the repair of those deteriorated concrete structures abroad contaminated by chloride ions [19, 20].

Considering the cost, two difficult choices are often encountered in practice for reinforcement protection in chloride-containing environment. The first is letting the structure be used normally for a period, then apply the cathodic protection system to it according to the corrosion state or degree of the reinforcement. The second is embedding the cathodic protection system in the new structure during construction, so it can be started immediately or wait for the right opportunity to start.

Two concepts are to be separated here, the corrosion "prevention" and the corrosion "protection" [20]. *The former refers to the preventive protection for the metals that have not been corroded, while the latter refers to the mandatory protection for the metals that have been corroded. The former requires a small cathodic protection*

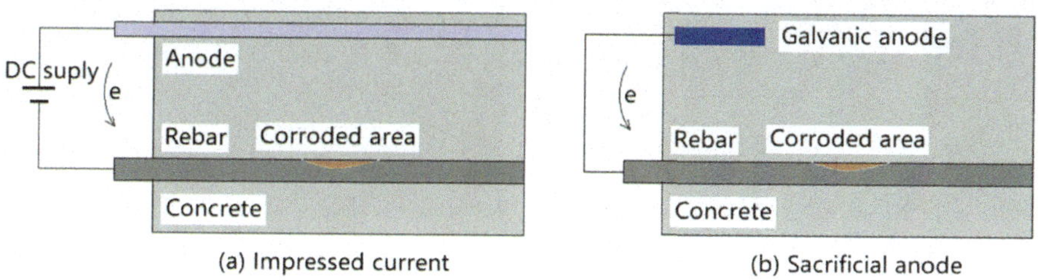

Fig. 8.2 Schematic of the two types of cathodic protection

current, while the latter requires a large cathodic protection current (usually several or dozens of times larger than the former).

For important engineering projects, the impressed current cathodic prevention or protection is usually the first choice. Many people believe that the sacrificial anode is more suitable for the protection of reinforcement in chloride contaminated concrete rather than in the newly built structures. They consider that the sacrificial anodes will be consumed prematurely and will not play any protective role in case of the rebar happens to corrode. In fact, this is related to the concrete structure and the layout of sacrificial anodes. In practice, the embedded sacrificial anodes are wrapped with specially designed porous mortars. The porous mortar can help the sacrificial anode be started in time and kept active when there is any corrosive medium penetrated. In addition, it can also accommodate the corrosion products of the sacrificial anode to avoid swelling and cracking caused by the direct contact with the concrete (different from the hot-dip galvanized reinforcement mentioned later).

From 2006 to 2008, we studied the effectiveness of distributed sacrificial anodes for the rebar in different strength concretes [21], and demonstrated the application in an expressway bridge deck in North China in 2007, which was subjected to deicing salt in winter. The impressed current and sacrificial anode system are both adopted in the demonstration project for comparison (see Figs. 8.3 and 8.4). The sacrificial anodes used are some imported from abroad and some are self-developed. Both the indoor research and field monitoring results show that, like the impressed current cathodic protection, the reasonably arranged sacrificial anodes can also provide good cathodic prevention for the black rebars.

Hot-dip galvanized rebars also use the sacrificial anodic protection principle, which was once popular and is still being promoted nowadays [22]. *As zinc is an amphoteric metal, it will corrode in concrete, and the volume expansion of corrosion products generated in chloride environment is about 2.6 times of the corroded zinc* [23], *it is recommended that the hot-dip galvanized rebars should better be used*

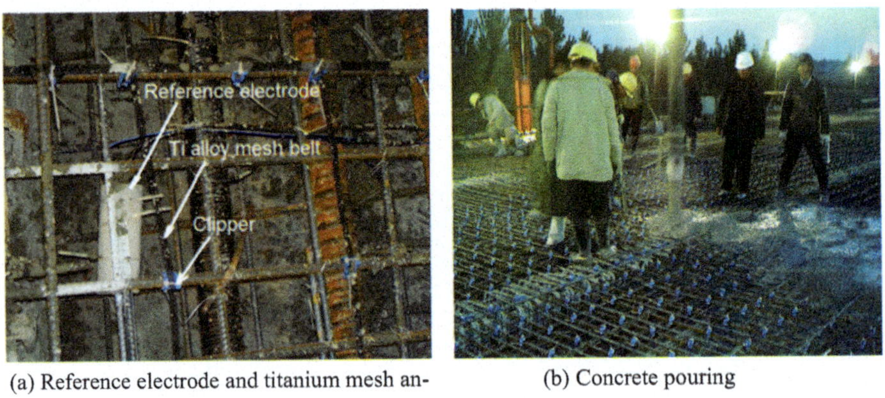

(a) Reference electrode and titanium mesh anode ribbon

(b) Concrete pouring

Fig. 8.3 Impressed current cathodic protection of reinforcement

(a) Reference electrode and sacrificial anode (b) Concrete pouring

Fig. 8.4 Sacrificial anode cathodic protection

in general corrosive environment (such as ordinary atmosphere or weak corrosive medium) rather than chloride containing environments.

8.4 Change the Rebar Interfacial Environment

Rebar corrosion inhibitor, electrochemical desalination, or re-alkalization are all methods to change the local environment at the concrete/rebar interface.

Corrosion inhibitors can affect the anodic or cathodic reaction of metals in corrosive media according to the vast results in labs, as shown in Fig. 8.5 for example [24]. So, it has been widely used in practice in the past decades. However, its long-term effect in engineering is often questioned, because it can also leach from the concrete into the wet environment. Therefore, ACI 222 suggests that it's better to be used in those concretes with low permeabilities and the water/cement ratio no more than 0.40 [8]. Unfortunately, there is not enough statistical data yet at present about the corrosion induction period delayed and the inhibition effect on the rebar corrosion propagation in practice.

Electrochemical desalination [25, 26] is aimed at the concrete polluted by the chloride ions. It uses the same charge repulsion principle to set up a positive electrode on the surface of water-saturated concrete and negatively charge the rebar, so that the chloride ions can migrate out of the concrete under electric field, see Fig. 8.6. Obviously, this technology is only applicable to those concretes with high permeabilities [27].

The electrochemical desalination and the impressed current cathodic protection are somewhat similar. But there are some differences indeed:

(1) ***The purpose is different***. *The electrochemical desalination is mainly used to change the local environment and slow down the corrosion of reinforcement;*

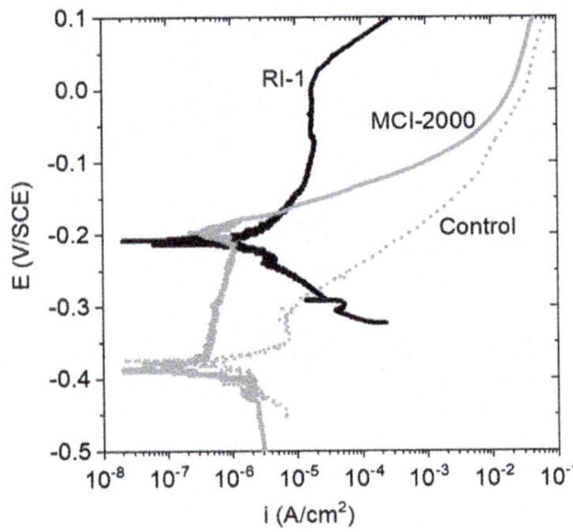

Fig. 8.5 Potentiodynamic polarization curves of polished HPB235 in 3% NaCl-Saturated $Ca(OH)_2$ + 3.6% RI-1 or 25 ml/L MCI-2000. RI-1 is a nitrite-based solid inhibitor. MCI-2000 is an alcohol amine-based liquid inhibitor [24]

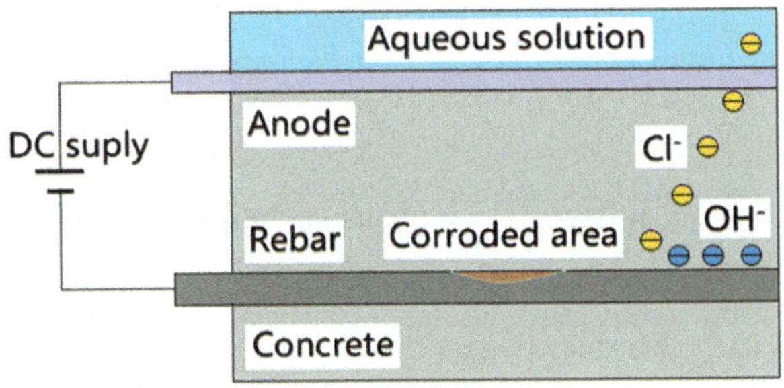

Fig. 8.6 Schematic of the electrochemical desalination and re-alkalization process

while the impressed current cathodic protection is to inhibit the corrosion of rebar;

(2) ***The current densities applied are different****. The electrochemical desalination usually uses the current density of 1–3 A/m^2, which is 2–3 orders of magnitude larger than that of the impressed current cathodic protection;*

(3) ***The devices are different****. The electrochemical desalination can use a simple constant current source or steady voltage supply, while the impressed current cathodic protection needs a precision constant current source, and usually requires a potential monitoring module; in addition, the electrochemical desalination requires setting the electrolyte solution on concrete surface, while the cathodic protection does not;*

(4) ***The anodes can be different****. Cheap and ordinary metals can be used for the electrochemical desalination, which can be corroded; but the impressed current cathodic protection needs highly corrosion-resistant materials as anode, such as titanium alloy, etc.;*

(5) ***The processes and requirements are different****. Except for the specific operation details, the duration for electrochemical desalination is short, usually several weeks to several months, while the impressed current cathodic protection is long-term, which might take at least several to dozens of years.*

During the electrochemical desalination process, the cathodic reaction around the reinforcement will increase the pH value of the concrete pore solution near the rebar surface. Therefore, people use this principle to treat the carbonized concrete and improve the pH value of the concrete pore solution, which is called electrochemical re-alkalization process, see Fig. 8.6.

Since the electrochemical desalination and the re-alkalization techniques use high currents, their negative effects have been often mentioned [28]. It is known from Sect. 6.4 that the reduction of mill scale can take place even without the hydrogen evolution when a negative current is applied, which may influence the bond strength of concrete to the rebar and the subsequent corrosion resistance. Therefore, for the actual hot-rolled black rebars with mill scale, especially for the corroded rebars, whether the electrochemical desalination or re-alkalization can be used needs to be carefully evaluated in specific situations.

8.5 Physical Isolation or Substitution

Increasing the thickness of concrete cover to prolong the induction time of rebar corrosion is not always the best choice, because too thick cover will significantly increase the cracking risk. Therefore, any surface modifications to protect the rebar or concrete surface are all welcome, which cut the aggressive media directly contact with the reinforcement, see Fig. 8.7 and Table 8.7.

Research in the United States indicates that [29, 30], compared with the ordinary reinforcement, the epoxy coated rebar can usually prolong the service life of the structure by 15–25 years. However, its usage is limited since high cost and it may reduce the bonding strength to concrete by about 20%, as well as its surface damage is not allowed in construction (more rapid local corrosion will occur). So, the epoxy coated rebar is often used as the first layer of the rebar mesh, as the stainless steel or galvanized rebars as well.

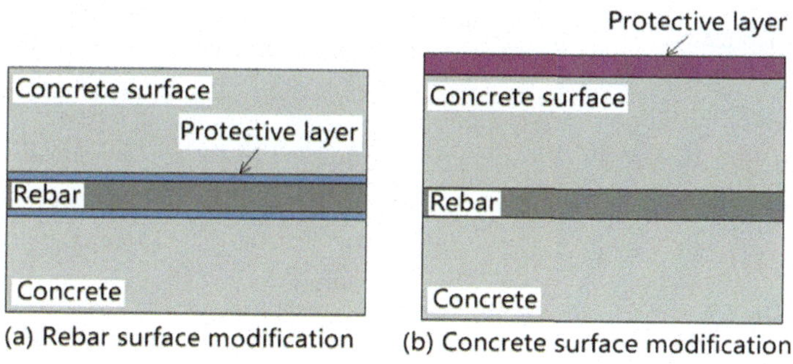

Fig. 8.7 Schematic of protection method through surface modification of rebar or concrete

Table 8.7 Modifications to rebar or concrete surface

Surface modified	Methods
Rebar surface	(1) Epoxy coating (2) Stainless steel cladding (3) Electroless nickel plating
Concrete surface	(1) Surface sealing: infiltration, organic/inorganic coating (2) Hydrophobic treatment: external coating or internal mixing of organic or inorganic hydrophobic agents such as silane (3) Corrosion-resistant cover/sheath: stainless steel, fibre reinforced polymer, composite materials, etc

The bond strength of polymer rebar to concrete is also lower than that of the ordinary rebars, its fire resistance is poor and the price is high, so it has not been widely used yet. Considering the technical and cost, among the various corrosion-resistant rebars mentioned above, the stainless-steel cladding ordinary carbon rebar might be a promising choice [31], but its connection details and on-site construction quality control are a bit of obstacles to expand its use.

Surface protective cover or sheath for concrete is not only an important protection approach for newly-built reinforced concretes, but also an important repair measure for the degraded concretes.

Table 8.8 lists some typical chloride diffusivities in different concretes obtained by the NEL method [32]. The permeability of ultra-high performance concrete (UHPC) [33] is two or three orders of magnitude lower than that of high performance concrete (HPC) or ordinary Portland concrete (OPC), which can effectively delay the penetration of water and harmful media such as chlorides into concrete and significantly prolong the service life of the structure. Therefore, besides the ultra-high strength, the outstanding impermeability is its truly great advantage. Metals in UHPC are supposed not to corrode since lack of water and oxygen owing to its ultra-low permeability.

Although UHPC has been studied almost near a half century, it has not been gradually expanded and attracted so extensive attentions until recently [34]. To make

Table 8.8 Chloride diffusivities in different concretes

Concrete	Compressive strength (MPa)	Chloride diffusivity ($\times 10^{-14} m^2/s$)
OPC	$\leq$10	$\geq$1000
	10–30	800–1000
HPC	30–60	500–800
	60–90	200–500
	90–120	100–200
UHPC	120–150	$\leq$20
	$\geq$150	$\leq$2

OPC–Ordinary Portland Concrete, HPC-High-performance Concrete, UHPC-Ultra-high-performance Concrete

full use of its ultra-high strength and impermeability, UHPC can be directly used as a powerful protective material for the deteriorated reinforced concrete structures. Of course, it can also combine with the normal concrete (NC) to form various UHPC-NC composite structures. While strengthening the conventional concretes, it can remarkably retard, mitigate, or even inhibit the rebar corrosion inside. So, it is one of the practical techniques that can be widely used at present and in the future.

8.6 Summary

The general principle and some protective methods for corrosion of hot-rolled black rebars are simply described in this chapter. Phosphating can be used to improve the atmospheric corrosion resistance, the cathodic protection can be chosen for the chloride contaminated concrete, and the UHPC is recommended for the protection of those deteriorated reinforced concretes.

References

1. Zhou, X., Nie, L., Hua, X., et al.: Influence of heat treatment temperature on structure and corrosion resistance of oxide scales formed on hot-rolled steel 510L. Trans. Mater. Heat Treat. **32**(10), 117–123 (2011). In Chinese
2. He, A.: The effect of oxide film on the electrochemical corrosion behaviors of hot rolled steel. Master thesis, Ocean University of China(2007). In Chinese.
3. Deng, J., Hu, J., Deng, P., et al.: Effect of oxide scales on initial corrosion behavior of SPHC hot rolled steel in tropical marine atmosphere. Chin. J. Chin. Soc. Corros. Protection. **39**(04), 331–337 (2019). In Chinese
4. Manna, M., Chakrabarti, I., Bandyopadhyay, N.: Phosphate treatment of TMT rebar bundle to avoid early rusting: An option for single step process. Surf. Coat. Technol. **201**, 1583–1588 (2006)
5. Wang, X.S.: High temperature fast phosphating and corrosion resistance test of hot-rolled rebar with mill scale. Master thesis, China University of mining and technology (2015)
6. Lu, X.Y., Wang, X.S., Li, Y.J., et al.: Preliminary study on high temperature rapid Zn-Ca recipe phosphating of rebar with mill scale. Suppl. Ind. Arch. **II**, 181–184 (2015). In Chinese
7. Lu, X.Y., Wang, X.S., Li, Y.J., et al.: Preliminary study on high temperature rapid Zn-Ni recipe phosphating of rebar with mill scale. Suppl. Ind. Arch. **II**, 185–188 (2015). In Chinese
8. Bremner, T., Hover, K., Poston, R., et al.: Protection of Metals in Concrete Against Corrosion, Technical Report for ACI Committee 222: Farmington Hills. MI, USA (2001)
9. NACE Standard RP-0169-96.: Control of External Corrosion on Underground or Submerged Metallic Piping Systems (1996)
10. Peabody, A.W., Bianchetti, R.L.: Control of Pipeline Corrosion, 2nd edn. NACE International, Housten (2001)
11. Wu, Y., Cao, B.: Cathodic Protection and Anodic Protection: Principle, Technology and Engineering Application. China Petrochemical Press, Beijing (2007)
12. Chess, P., Grønvold, F., Karnov, A.S.: Cathodic Protection of Steel in Concrete. E & FN Spon, London (1998)
13. Corrosion Protection: Concrete Bridges Publication No. FHWA-RD-98-088 (1998)

14. Tettamanti, M., Rossini, A.: Cathodic prevention and protection of concrete elements at the sydney opera house. Mater. Performance **36**(9), 21–25 (1997)
15. Clemeña, G.G., Pritchett, M.B., Napier, C.S.: Application of Cathodic Prevention in a New Concrete Bridge Deck in Virginia. VTRC 03-R11 (2003)
16. Hong, D., Fan, W., Luo, D., et al.: Preliminary study on impressed current cathodic protection technology for marine concrete superstructure. Corros. & Prot. **11**(5), 257–261 (1990). In Chinese
17. Hong, D.: Corrosion and Protection of Reinforcement in Concrete. China Railway Press, Beijing (1998).In Chinese
18. Ge,Y., Zhu, X.: Corrosion and Cathodic Protection of Reinforcement in Concrete Beijing, Chemical Industry Press (2007). In Chinese
19. Wang, W., Ji, M.: Research progress of sacrificial anode protection for reinforced concrete structures. Mar. Sci. **05**, 18–20+33 (2001). In Chinese
20. Pedeferri, P.: Cathodic protection and cathodic prevention. Constr. Build. Mater. **10**(5), 391–402 (1996)
21. Zhang, G.: Preliminary study on cathodic protection of distributed sacrificial anode on reinforcement in concrete. Master thesis, Tsinghua University (2008). In Chinese
22. Yeomans, S.R.: Galvanized Steel Reinforcement in Concrete. Elsevier Science, Amsterdam (2004)
23. Hime, W., Machin, M.: Performance variances of galvanized steel in mortar and concrete. Corrosion **49**(10), 858–860 (1993)
24. Lin, W.: Influences on rebar corrosion by rebar surface states, pH of concrete pore solution and [Cl^-] in concrete. Master Thesis, Tsinghua University (2002). In Chinese
25. Mietz, J.: Electrochemical Rehabilitation Methods for Reinforced Concrete Structures-A State of the Art Report, Leeds: The Institute of Materials Minerals and Mining (1998)
26. Poursaee, A.: Corrosion of Steel in Concrete Structures. Elsevier, Woodhead (2016)
27. Liu, X.: Study on electrochemical desalination of concrete of Tsinghua University, Master thesis (2002). In Chinese
28. Bastidas, D.M., Cobo, A., Otero, E., et al.: Electrochemical rehabilitation methods for reinforced concrete structures: advantages and pitfalls. Corros. Eng., Sci. Technol. **43**(3), 248–255 (2008)
29. Sagüés, A.A., Lau, K., Powers, R.G., et al.: Corrosion of epoxy-coated rebar in marine bridges-part 1: A 30-year perspective. Corrosion **66**(6), 065001-1–065001-13 (2010)
30. Lau,K., Sagüés, A.A., Powers, R.G.: Corrosion of epoxy-coated rebar in marine bridges-part 2: corrosion in cracked concrete. Corrosion **66**(6), 065002-1–065002-16 (2010)
31. Cui, F.S.: Corrosion Behavior of Stainless Steel Clad Rebar. University of South Florida (2003)
32. Lu, X.Y.: Application of the Nernst-Einstein equation to concrete. Cem. Concr. Res. **27**(2), 293–302 (1997)
33. Fehling, E., Schmidt, M., Walraven, J., et al.: Ultra-High Performance Concrete UHPC: Fundamentals, Design, Examples. Wilhelm Ernst & Sohn, Berlin (2014)
34. Zhao, J., Shi, H.X., Lu, X.Y.: Basic properties and test methods of ultra-high performance concrete. China Building Materials Industry Press (2019). In Chinese

Chapter 9
Further Work Needs to Do

Abstract The deficiencies in this monograph and some considerations about the future work needs to do are shortly described in this chapter. More attentions need to pay to the corrosion and protection of high-strength and the prestressing rebars, and to develop novel composite structures to improve the durability of the reinforced concretes.

9.1 Carriers' Problem

Because it is difficult to obtain the passivation and corrosion behavior of hot-rolled rebar with mill scale under very low dissolved oxygen concentrations in aqueous solution, the residual oxygen is still high enough relative to the requirement for the oxidation of iron under conventional deaeration conditions, i.e., the iron oxide can remain stable under these conditions. If further quantitative studies are demanded about the main charge carrier and its migration rate in the mill scale of hot-rolled rebar at low dissolved oxygen concentrations, it needs to carry out more refined experiments.

Traditionally, it is considered that the migration rate of Fe^{2+} and Fe^{3+} in Fe_3O_4 is low at room temperature. But the hopping of electrons between them is not too difficult [1]. This role in and influence on the Mott-Schottky curves and the redox at the metal/oxide or the oxide/solution interfaces can worthy be revisited in detail. Of course, the reactions within the mill scale, the contribution and migrating kinetics of oxygen vacancies and metallic ions also need to be carefully re-investigated in combination with the quantum chemical calculations [1–3], which may provide supporting proof to clarify what is the main carrier during the different redox processes.

The redox of magnetite [4–7] *and the reduction of hematite* [8, 9] *have been recently investigated using the density functional theory, and it is verified that the transfer of cation vacancies and interstitial iron atoms, and the electron hopping between the* $Fe_{oct}{}^{2+}$ *and* $Fe_{oct}{}^{3+}$ *are playing an important role during the redox reactions of magnetite. The change and diffusion of oxygen vacancies on the hematite surface may influence its reduction process and phase transformation. Some research*

X. Lu, *Passivation and Corrosion of Black Rebar with Mill Scale*, Engineering Materials,
https://doi.org/10.1007/978-981-19-8102-9_9

[10, 11] *on the electrode materials may help to understand the electrode processes and the phase transformations as well as the particle transfer involved.*

9.2 Influence of Adsorption Films

Since the surface of mill scale is rich in oxygen atoms, the research on oxygen adsorption film has not been carried out. If it is interested, further research can be conducted with the help of oxygen isotopes as tracers [12, 13]. Or it can be referenced to the literature about the oxidation of magnetite based on the first-principle calculations [1, 4–6].

This monograph does not concern the adsorption kinetics of corrosion inhibitors on hot-rolled rebar with mill scale. Our previous research was also based on the descaled rebars. If the protective effect of corrosion inhibitors [14], especially the organic ones, on the rebars with mill scale is concerned, practical tests, as well as quantum chemical calculations and molecular dynamic simulations are highly recommended [15, 16].

9.3 Influence of Stress

Both tensile and fatigue stresses can accelerate the metallic corrosion and its propagation, which may be more sensitive to the hot-rolled rebars with mill scale. The case shown in Fig. 9.1 is very common on the construction site. That means, after

Fig. 9.1 The bent is sensitive to corrosion[1]

(a) Corrosion before anchor sealing

(b) Corrosion in the duct

Fig. 9.2 Pre-rust and corrosion of strands on-site

the rebar with mill scale is tensioned, the number of cracks and debonding places in the mill scale will increase significantly, which can promote the rebar to rust earlier.

In fact, civil engineers still lack common sense about the stress corrosion cracking of the black rebars. Now the metallurgical and civil engineering industry are both vigorously promoting the high-strength reinforcements, while the corrosion resistance of the newly developed high-strength rebars has not been significantly improved comparing with the ordinary reinforcements. Therefore, the stress corrosion cracking of high-strength rebars needs to pay more attention in the future.

In addition, the problem of metallic corrosion in the whole prestressed systems has not been fully recognized and gotten enough attentions yet, especially on the corrosion of the prestressed strands [17], see Fig. 9.2.

Figure 9.3 is the pictures and the corrosion product morphologies of corroded prestressed steel strands taken from a demolished structure serviced less than 20 years [18]. The section Raman spectra analysis indicates that the outer corrosion product is mainly composed of α-FeOOH, γ-FeOOH, and α-Fe_2O_3 together with some $CaCO_3$, the middle part is α-FeOOH and γ-Fe_2O_3, and the inner is α-Fe_2O_3 and Fe_3O_4.

The maximum weight loss of 6 tested corroded steel strands is about 5.2%, and the minimum residual tensile strength is 1910 MPa which still meets the minimum requirements of 1860 MPa in the specification. However, the minimum yield strength is 1660 MPa, which is slightly lower than the minimum limit of 1670 MPa, and the elongation of 5 in 6 steel strands is lower than the minimum limit of 3.5% together with one of them has been reduced to less than 80% of the requirement.

According to our observations, the early stress corrosion cracking might be associated with the premature rusting of the strands (see Fig. 9.4) as well as the grouting defects.

To better understand the mechanisms of the stress corrosion cracking, the density functional theory has also been employed to the hydrogen embrittlement [19], *which*

[1] Photos in Figs. 9.1, 9.2, 9.3a and 9.4 on-site are taken by Prof. Baolin Guo.

(a) Prestressed strands

(b) Piece of corroded strand

(c) Surface morphology

(d) Section morphology

Fig. 9.3 Corroded prestressed steel strand and the morphology of corrosion product [18]

(a) Rust on fresh steel wires

(b) Rust on steel wires

Fig. 9.4 Premature rusting of steel wires

is quite sensitive to the high strength rebar or prestressed wires, as well as the chloride permeation through the oxide scale [20]. *So, the first-principle method is a good tool for corrosion investigation, and it is welcome in the future to rebars' territory.*

As concrete technology is running into the UHPC times, some traditional problems, such as the corrosion of rebar may be changed or totally solved in the future, even though new problems may come. Then, how to systematically design novel concrete structures and how to protect them from rapid deteriorations should be carefully concerned. In short, there is still a lot of work needs to do.

References

1. Parkinson, G.S.: Iron oxide surfaces. Surf. Sci. Rep. **71**(1), 272–365 (2016)
2. Meng, Y., Liu, X.W., Huo, C.F., et al.: When density functional approximations meet iron oxides. J. Chem. Theory Comput. **12**(10), 5132–5144 (2016)
3. Liao, P.: Mechanical, optical, transport, and catalytic properties of iron oxides from first principles. Ph.D. dissertation, Princeton University (2012)
4. Santos-Carballal, D., Roldan, A., Grau-Crespo, R., et al.: A DFT study of the structures, stabilities and redox behaviour of the major surfaces of magnetite Fe_3O_4. Phys. Chem. Chem. Phys. **16**, 21082–21097 (2014)
5. Nie, S., Starodub, E., Monti, M., et al.: Insight into magnetite's redox catalysis from observing surface morphology during oxidation. J. Am. Chem. Soc. **135**, 10091–10098 (2013)
6. Kendelewicz, T., Kaya, S., Newberg, J.T., et al.: X-ray photoemission and density functional theory study of the interaction of water vapor with the Fe_3O_4 (001) surface at near-ambient conditions. J. Phys. Chem. C **117**(6), 2719–2733 (2013)
7. Mulakaluri, N., Pentcheva, R.: Hydrogen adsorption and site-selective reduction of the Fe_3O_4 (001) surface: insights from first principles. J. Phys. Chem. C **116**(31), 16447–16453 (2012)
8. Schöttner, L., Nefedov, A., Yang, C., et al.: Structural evolution of α-Fe_2O_3 (0001) surfaces under reduction conditions monitored by infrared spectroscopy. Front. Chem. **451** (2019)
9. Zhu, W., Winterstein, J., Maimon, I., et al.: Atomic structural evolution during the reduction of α-Fe_2O_3 nanowires. J. Phys. Chem. C **120**(27), 14854–14862 (2016)
10. Lininger, C.N.: The mechanistic description of the open circuit potential for the lithiation of magnetite nanoparticles. Ph.D. dissertation, Columbia University (2018)
11. Zhang, W., Bock, D.C., Pelliccione, C.J., et al.: Insights into ionic transport and structural changes in magnetite during multiple electron transfer reactions. Adv. Energy Mater. **6**(10), 1502471 (2016)
12. Bardwell, J.A., Macdougall, B., Graham, M.J.: Use of ^{18}O/SIMS and electrochemical techniques to study the reduction and breakdown of passive oxide films on iron. J. Electrochem. Soc. **135**(2), 413–418 (1988)
13. Wang, L., Voyshnis, S., Seyeux, A., et al.: Ion transport mechanisms in the passive film formed on 304L stainless steel studied by ToF-SIMS with ^{18}O isotopic tracer. Corros. Sci. **173**, 108779 (2020)
14. McCafferty, E.: Introduction to Corrosion Science, pp. 357–402. Springer, New York (2010)
15. Noh, J.H.: First-principles theoretical characterization of the electronic structure of iron and iron oxide surfaces and their interfaces with organic layers. Ph.D. Dissertation, Georgia Institute of Technology (2017)
16. Taylor, C.D., Marcus, P.: Molecular Modeling of Corrosion Processes: Scientific Development and Engineering Applications. John Wiley & Sons, New Jersey (2015)
17. ACI 222.2R-14.: Report on Corrosion of Prestressing Steels. American Concrete Institute (2014)

18. Jiang, K.: Corrosion Analysis and simulation test of strands embedded in duct grout of an Offshore Bridge. Tsinghua University, Master thesis (2013). In Chinese
19. Chohan, U.K.: Modelling early stages of hydrogen embrittlement and surface oxidation of iron using density functional theory. Ph.D. Dissertation, The University of Manchester, United Kingdom (2018)
20. Cantatore, V., Olivas Ogaz, M.A., Liske, J., et al.: Oxidation driven permeation of Iron oxide scales by chloride from experiment guided first-principles modeling. J. Phys. Chem. C **123**(42), 25957–25966 (2019)

Epilogue

When I was just in primary school, my father told me a story about an ancient lawyer who changed the word "killing with a knife" to "killing with a thrown knife"[1] in the file and saved a spoiled boy. He told me that "if a character is not written in the right way, it can harm others and even yourself. You should not be careless and casual when you write!".

The story must have been made up by himself, but since then, whenever I pick up the pen to write, I always recall the story. Like this monograph, in the past ten years, I started to write several times but stopped soon. Over the years, it has become an excuse for my laziness. Last time, I spent lots of efforts, wrote more than a half, but finally gave it up. Because I felt that something was inconclusive, and there are so many monographs already, it seems not necessary to write more lines.

At the end of last year, my daughter recommended to me a "Book of why: the new science of cause and effect" written by Judea Pearl and Dana Mackenzie. I found the Chinese version, read it over, and got inspired. So, I decided to complete the monograph.

This Spring Festival, due to epidemic control, we three generations are isolated in three different places. Plus, due to the Beijing Winter Olympics, I can only stay at home, so I turn over the past manuscripts and cut them back and forth, which has become a strange style as you have seen now.

The original title of the book was intended to be: Is there a passive film on the actual black rebar? But finally, it was changed to the current one. You may see the traces of my tossing back and forth between them.

Looking back to the past, my undergraduate thesis was about the broken mechanisms of passive film on 304 stainless steel. Later, I turned to civil engineering and encountered the problem of rebar corrosion. I don't know if this is my fate or if there is any "cause and effect?" From Professor Pearl's view, there might be any causality involved.

---

[1] Chinese "use" and "throw" have minor difference in writing.

X. Lu, *Passivation and Corrosion of Black Rebar with Mill Scale*, Engineering Materials, https://doi.org/10.1007/978-981-19-8102-9

Due to my lack of knowledge, I haven't made it clear as expected about the rebar corrosion during the past more than 20 years, causing this monograph so wordy that may not be clear, too. What I want to say is that the actual hot-rolled black rebars with mill scale in concrete might not produce the "legendry" passive film, rebar corrosion occurs preferentially at the defects of the mill scale, and where the pre-rusted stains, which is difficult to be re-passivated and ready to corrode all the time. Of course, this monograph does not intend to deny the previous conclusions, but can be another version of "a blind man touching an elephant".

However, from the perspective of the durability of reinforced concrete, if it can be recognized that the actual black rebar cannot generate a passive film, it will be helpful to improve the durability design and take different protective measures for reinforced concretes at present or in the future.

In the mill plant, high-temperature phosphating and any other better anti-corrosion treatments for black rebars are suggested to be carried out first, so the hot-rolled rebars can be free from rusting before the concrete placing. Sealing the concrete surface or other methods that can significantly reduce the permeability of the concrete are particularly recommended. Repairing and strengthening the deteriorated reinforced concrete in a harsh environment with UHPC is another economic and long-term method I prefer.

When I finished writing, I still recalled the story my father told. Here, I use this monograph to pay my sincere tribute to my parents and relatives, and to those who love me and the people I love in! I hope the words in this monograph will not confuse readers more. Any corrections and comments are all welcome!

Appendix A
Chemical Stability of Metals

Categories of metals according to their standard hydrogen electrode potentials are listed in Table A.1.

Table A.1 Categories of metals by their standard hydrogen potentials[a]

Group	Chemical stability	Electrode reaction	Potential(V/SHE)
I	Very active, base metals Can corrode in neutral aqueous medium without oxygen or oxidizers	$Li \rightleftharpoons Li^{+} - e$	−3.045
		$Rb \rightleftharpoons Rb^{+} - e$	−2.925
		$K \rightleftharpoons K^{+} - e$	−2.925
		$Cs \rightleftharpoons Cs^{+} - e$	−2.923
		$Ra \rightleftharpoons Ra^{+} - e$	−2.92
		$Ba \rightleftharpoons Ba^{+} - e$	−2.9
		$Sr \rightleftharpoons Sr^{+} - e$	−2.89
		$Ca \rightleftharpoons Ca^{+} - 2e$	−2.87
		$Na \rightleftharpoons Na^{+} - e$	−2.714
		$La \rightleftharpoons La^{+} - 3e$	−2.52
		$Mg \rightleftharpoons Mg^{2+} - 2e$	−2.37
		$Am \rightleftharpoons Am^{3+} - 3e$	−2.32
		$Pu \rightleftharpoons Pu^{3+} - 3e$	−2.07
		$Th \rightleftharpoons Th^{4+} - 4e$	−1.9
		$Np \rightleftharpoons Np^{3+} - 3e$	−1.86
		$Be \rightleftharpoons Be^{2+} - 2e$	−1.85
		$V \rightleftharpoons V^{3+} - 3e$	−1.8
		$Hf \rightleftharpoons Hf^{4+} - 4e$	−1.7
		$Al \rightleftharpoons Al^{3+} - 3e$	−1.66
		$Ti \rightleftharpoons Ti^{2+} - 2e$	−1.63
		$Zr \rightleftharpoons Zr^{4+} - 4e$	−1.53

(continued)

X. Lu, *Passivation and Corrosion of Black Rebar with Mill Scale*, Engineering Materials, https://doi.org/10.1007/978-981-19-8102-9

Table A.1 (continued)

Group	Chemical stability	Electrode reaction	Potential(V/SHE)
		$Ti \rightleftharpoons Ti^{3+} - 3e$	−1.21
		$V \rightleftharpoons V^{2+} - 2e$	−1.18
		$Mn \rightleftharpoons Mn^{2+} - 2e$	−1.18
		$Nb \rightleftharpoons Nb^{3+} - 3e$	−1.1
		$Ta \rightleftharpoons Ta^{2.5+} - 2.5e$ (Ta_2O_5)	−0.81
		$Zn \rightleftharpoons Zn^{2+} - 2e$	−0.762
		$Cr \rightleftharpoons Cr^{3+} - 3e$	−0.74
		$Ga \rightleftharpoons Ga^{3+} - 3e$	−0.53
		$Fe \rightleftharpoons Fe^{2+} - 2e$	−0.44
II	Active, base metals Stable in oxygen free neutral medium and can corrode in oxygen free acidic media	$Cd \rightleftharpoons Cd^{2+} - 2e$	−0.402
		$In \rightleftharpoons In^{3+} - 3e$	−0.342
		$Tl \rightleftharpoons Tl^{3+} - 3e$	−0.336
		$Mn \rightleftharpoons Mn^{3+}$-3e	−0.283
		$Co \rightleftharpoons Co^{2+} - 2e$	−0.277
		$Ni \rightleftharpoons Ni^{2+} - 2e$	−0.25
		$Mo \rightleftharpoons Mo^{3+} - 3e$	−0.2
		$Ge \rightleftharpoons Ge^{4+} - 4e$	−0.15
		$Sn \rightleftharpoons Sn^{2+} - 2e$	−0.136
		$Pb \rightleftharpoons Pb^{2+} - 2e$	−0.126
		$W \rightleftharpoons W^{3+} - 3e$	−0.11
III	Semi-noble metals Stable in acidic and neutral medium without oxygen and oxidizers	$Bi \rightleftharpoons Bi^{3+} - 3e$	+0.226
		$Sb \rightleftharpoons Sb^{3+} - 3e$	+0.24
		$Re \rightleftharpoons Re^{3+} - 3e$	+0.30
		$As \rightleftharpoons As^{3+} - 3e$	+0.30
		$Cu \rightleftharpoons Cu^{2+} - 2e$	+0.337
		$Hg \rightleftharpoons Hg^{+} - e$	+0.789
		$Ag \rightleftharpoons Ag^{+} - e$	+0.799
		$Rh \rightleftharpoons Rh^{3+} - 3e$	+0.80
IV	Inert, noble metals Don't corrode in aerobic neutral medium and corrode in acidic medium with oxygen or oxidizers	$Pd \rightleftharpoons Pd^{2+} - 2e$	+0.987
		$Ir \rightleftharpoons Ir^{3+} - 3e$	+1.000
		$Pt \rightleftharpoons Pt^{2+} - 2e$	+1.19
V	Completely inert, noble metals Stable in acidic medium with oxygen but can dissolve in complex-forming medium in the presence of oxidizers	$Au \rightleftharpoons Au^{3+} - 3e$	+1.5

[a] Reprinted with permission of Springer Nature from: Tomashov ND, Chernova GP (1967) Passivity and Protection of Metals Against Corrosion. New York: Plenum Press. p2

Appendix B
Iron Compounds

The composition and physical properties of some iron compounds are listed in Table B.1.

X. Lu, *Passivation and Corrosion of Black Rebar with Mill Scale*, Engineering Materials,
https://doi.org/10.1007/978-981-19-8102-9

Table B.1 Composition and physical properties of common iron compounds[a]

Formula	Name	Color	Density (g/cm^3)	Crystal structure	Cell dimension (nm)	Magnetism	ΔG_f^o (kJ/mol)
FeO	Wüstite	Black	5.9–5.99	Cubic	a=0.4275−0.4302	Antiferromag	−251
Fe_3O_4	Magnetite	Black	5.18	Cubic	a=0.8396	Ferrimag	−1012.6
$\gamma-Fe_2O_3$	Maghemite	Reddish – brown	4.87	Cubic or Tetragonal	a=0.83474	Ferrimag	−711.1
α-Fe_2O_3	Hematite	Red	5.26	Rhombohedral	a=0.50356(1)	Weakly Ferromag. or Antiferromag	−742.7
				Hexagonal	c=1.37489(7)		
$\alpha-FeOOH$	Goethite	Yellow–brown	4.26	Orthorhombic	a=0.9956 b=0.30215	Antiferromag	−488.6
					c=0.4608		
β-FeOOH	Akaganéite	Yellow–brown	–	Monoclinic	a=1.0546 b=0.3031 c=1.0483 β=90.63°	Antiferromag	–
γ-FeOOH	Lepidocrocite	Orange	4.09	Orthorhombic	a=0.307 b=1.253 c=0.388	Antiferromag	−477.7

(continued)

Table B.1 (continued)

Formula	Name	Color	Density (g/cm^3)	Crystal structure	Cell dimension (nm)	Magnetism	ΔG_f^o (kJ/mol)
δ'−FeOOH	Feroxyhyte	Red-brown	4.2	Hexagonal	a=0.293 c=0.456	Ferrimag	–
$Fe(OH)_2$	Ferrous hydroxide	White	3.4	Hexagonal	a=0.3258 c=0.4605	–	–
$Fe(OH)_3$	Bernalite	Dark green	3.32	Orthorhombic	a=0.7544(2) b=0.7560(4) c=0.7558(2)	–	–
$Fe_5HO_8{\cdot}4H_2O$	Ferrihydrite	Red-brown	3.96	Hexagonal	a=0.2955 c=0.937	Speromagnet	−699
$Fe_{16}O_{16}(OH)_y(SO_4)_z{\cdot}nH_2O$	Schwertmannite	Orange−brown	3.8	Tetragonal	a=1.065 c=0.604	Antiferromag	–

[a] Reprinted with permission of Wiley–VCH from: Cornell RM, Schwertmann U (2003) The iron oxides: structure, properties, reactions, occurences and uses. WILEY–VCH Verlag GmbH & Co. Weinheim. p4-5

Appendix C
Experimental Materials

1. **Composition of rebar**

The elemental compositions of rebar matrix are shown in Table C.1, which is examined by the analysis and inspection center of University of Science and Technology Beijing.

2. **Preparation of specimens**

The preparation methods of rebar samples used in this monograph are listed in Table C.2.

Pickling process for descaling mill scale: mix concentrated HCl with distilled water in the volume ratio of 1:3 and add 0.5% hexamethylenetetramine ($C_6H_{12}N_4$) to form pickling solution. Dip the hot-rolled rebar in the pickling solution at room temperature for 30 min, then rinse it with absolute ethanol, and ultrasonically clean it for 10 min, and finally dry it with an electric hair dryer for preparation.

The "matrix-mill scale galvanic couple" sample of HRB335 rebar is shown in Fig. C.1.

Photos of HRB400 rebar and mill scale samples are shown in Fig. C.2.

Table C.1 The diameter and matrix element composition of rebar

Grade	Diameter (mm)	Chemical element content (%wt)					
		C	Si	Mn	S	P	Fe
HPB235[a]	8	0.18	0.19	0.52	0.017	0.017	Balanced
HRB335	12	0.22	0.38	0.48	0.031	0.030	
HRB400	12	0.31	0.23	0.38	0.012	0.021	

[a] Once called as A_3 or Q235

X. Lu, *Passivation and Corrosion of Black Rebar with Mill Scale*, Engineering Materials, https://doi.org/10.1007/978-981-19-8102-9

Table C.2 Sample preparation methods

Rebar	Surface state	Area tested	Preparation
HPB235	Polished	Section	Sandpaper grinding, metallographic polishing; The remaining surface is sealed with 914 epoxy resin; The test surface shall be cleaned with absolute ethanol and acetone
	Grounded	Longitudinal surface	Sandpaper grinding, surface roughness < 1.6 μm; Both ends are sealed with 914 epoxy resin
	With mill scale		The surface is cleaned with absolute ethanol and acetone, and the exposed surface length is about 20 mm
	With rust stains		
HRB335	Polished	Section	Descaled by pickling and ground using sandpaper to 1000#; The remaining surface is sealed with 914 epoxy resin; The test surface is cleaned with absolute ethanol, the same below
	Descaled	Longitudinal surface	Descaled by pickling, both ends are sealed with 914 epoxy resin, and the exposed surface length is about 20 mm
	With mill scale		Both ends are sealed with 914 epoxy resin, and the exposed surface length is about 20 mm
HRB400	Descaled	Longitudinal surface	Descaled by steel wheel, both ends are sealed by epoxy resin, with about 20 mm exposed in the middle; Ultrasonic cleaning in absolute ethanol for 10 min, the same below
	Matrix	Section	Descaled by steel wheel, ground using sandpaper to 2000#; The remaining surface is sealed with epoxy resin, the edge of working face is sealed with 703 silicone rubber, and the exposed area is about 0.5 cm^2
	With mill scale	Longitudinal surface	Both ends are sealed with epoxy resin, and the exposed surface length is about 20 mm in the middle
	Mill scale	Inner surface	Peeled off from rebar surface using epoxy resin; Coat one end with silver conductive adhesive, lead the copper conductor and seal it with epoxy, and the test surface is about 0.5 cm^2

3. **Solutions**

Simulated concrete pore solutions are listed in Table C.3.

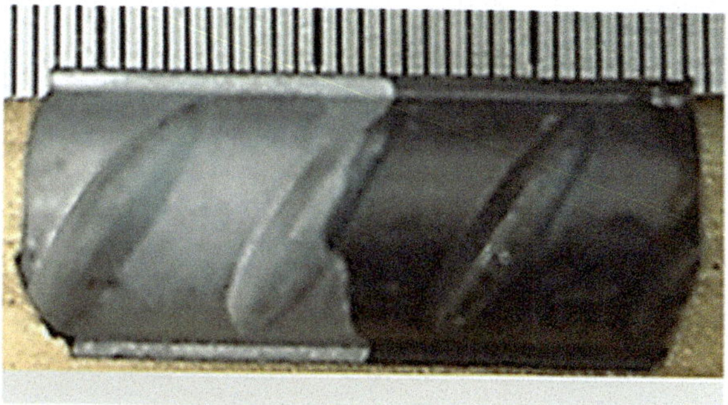

Fig. C.1 Matrix-mill scale galvanic couple sample of HRB335 rebar

(a) with mill scale

(b) descaled

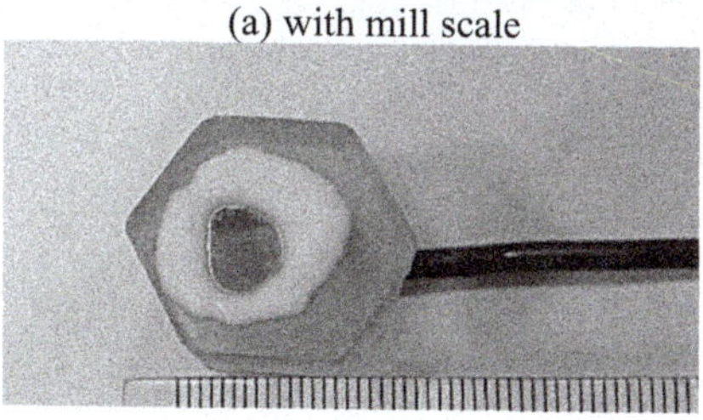

(c) matrix

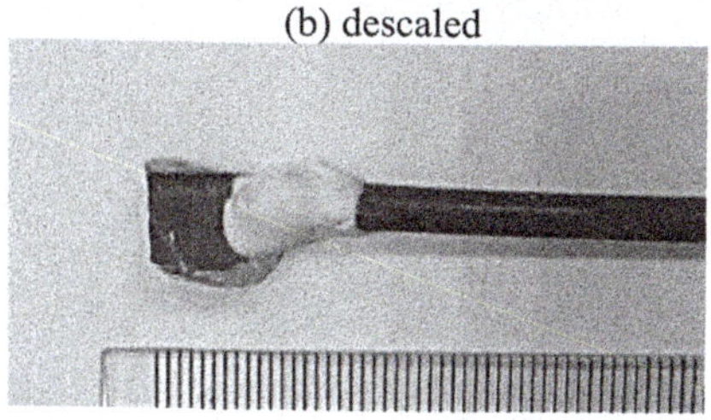

(d) mill scale

Fig. C.2 HRB400 rebar with mill scale, descaled, rebar matrix, and mill scale samples

Table C.3 Composition, pH, and conductivity of simulated concrete pore solutions

Composition	pH	Conductivity (mS/cm)	Dissolved oxygen (mg/L)
Saturated $Ca(OH)_2$	12.5–12.6	8.03–8.08	8.04
0.6 mol/L KOH + 0.2 mol/L NaOH + Saturated $Ca(OH)_2$[a]	10.0–13.5	/	/
18 g/L KOH + 5.2 g/L NaOH + Saturated $Ca(OH)_2$	13.6	60.04	7.46
Saturated $Ca(OH)_2$ + 5 ppm NaCl	12.5	8.24	8.30

[a] This is for HPB235 tests, the others for HRB335 tests

Index

X. Lu, *Passivation and Corrosion of Black Rebar with Mill Scale*, Engineering Materials,
https://doi.org/10.1007/978-981-19-8102-9

The manufacturer's authorised representative in the EU is Springer Nature Customer Service Centre GmbH, Europaplatz 3, 69115 Heidelberg, Germany. If you have any concerns regarding our products, please contact ProductSafety@springernature.com

Printed and bound by CPI Group (UK) Ltd, Croydon, CR0 4YY
30/09/2026
02225308-0006